P9-AQA-362

# CENTRE AND PERIPHERY

# SAGE FOCUS EDITIONS

1. **POLICE AND SOCIETY,** edited by David H. Bayley
2. **WHY NATIONS ACT: Theoretical Perspectives for Comparative Foreign Policy Studies,** edited by Maurice A. East, Stephen A. Salmore, and Charles F. Hermann
3. **EVALUATION RESEARCH METHODS: A Basic Guide,** edited by Leonard Rutman
4. **SOCIAL SCIENTISTS AS ADVOCATES: Views from the Applied Disciplines,** edited by George H. Weber and George J. McCall
5. **DYNAMICS OF GROUP DECISIONS,** edited by Hermann Brandstatter, James H. Davis, and Heinz Schuler
6. **NATURAL ORDER: Historical Studies of Scientific Culture,** edited by Barry Barnes and Steven Shapin
7. **CONFLICT AND CONTROL: Challenge to Legitimacy of Modern Governments,** edited by Arthur J. Vidich and Ronald M. Glassman
8. **CONTROVERSY: Politics of Technical Decisions,** edited by Dorothy Nelkin
9. **BATTERED WOMEN,** edited by Donna M. Moore
10. **CRIMINOLOGY: New Concerns,** edited by Edward Sagarin
11. **TO AUGUR WELL: Early Warning Indicators in World Politics,** edited by J. David Singer and Michael D. Wallace
12. **IMPROVING EVALUATIONS,** edited by Lois-ellin Datta and Robert Perloff
13. **IMAGES OF INFORMATION: Still Photography in the Social Sciences,** edited by Jon Wagner
14. **CHURCHES AND POLITICS IN LATIN AMERICA,** edited by Daniel H. Levine
15. **EDUCATIONAL TESTING AND EVALUATION: Design, Analysis, and Policy,** edited by Eva L. Baker and Edys S. Quellmalz
16. **IMPROVING POLICY ANALYSIS,** edited by Stuart S. Nagel
17. **POWER STRUCTURE RESEARCH,** edited by G. William Domhoff
18. **AGING AND SOCIETY: Current Research Perspectives,** edited by Edgar F. Borgatta and Neil G. McCluskey
19. **CENTRE AND PERIPHERY: Spatial Variation in Politics,** edited by Jean Gottmann
20. **THE ELECTORATE RECONSIDERED,** edited by John C. Pierce and John L. Sullivan

# CENTRE AND PERIPHERY

## SPATIAL VARIATION IN POLITICS

Edited by

**Jean Gottmann**

*Published with the cooperation of the Maison des Sciences de l'Homme and the Permanent Research Committee on Political Geography of the International Political Science Association*

*For information address:*

SAGE Publications
275 South Beverly Drive
Beverly Hills, California 90212

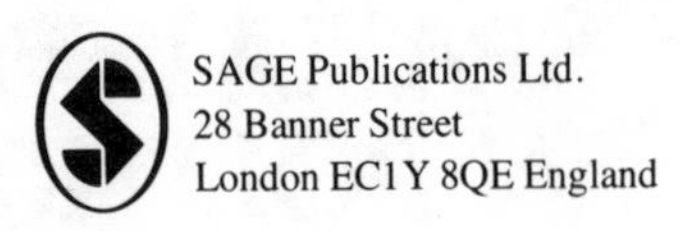

SAGE Publications Ltd.
28 Banner Street
London EC1Y 8QE England

Printed in the United States of America

**Library of Congress Cataloging in Publication Data**

Main entry under title:

Centre and periphery.

(Sage focus editions; 19)
"The theme of Center and Periphery was adopted for a symposium held by the Committee on Political Geography of the International Political Science Association in Paris, in January 1978 . . . based on papers offered at that meeting."
1. Geography, Political—Congresses. I. Gottmann, Jean. II. Series.
JC319.C45 320.9 79-21564
ISBN 0-8039-1344-3
ISBN 0-8039-1345-1 pbk.

FIRST PRINTING

## CONTENTS

# PREFACE

Why "centre and periphery" now, and by such a diverse group of scholars? The choice of this theme may well be a sign of the times. In an era of rapid change in socioeconomic structures, political science and political geography both look for new tools of analysis, and for systems of relationships that would be endowed with permanent value and that would provide a better understanding of the dynamic evolution affecting many of the spatial patterns with which the political process is concerned.

The theme of centre and periphery was adopted for a symposium held by the Committee on Political Geography of the International Political Science Association in Paris, January 1978. The chapters that follow in this book are based on the papers offered at that meeting and on the ideas that emerged from the discussion of these papers.

Perhaps I should recall here the sequence of debates that gave birth to this book. The International Political Science Association had adopted for the deliberations of its Tenth World Congress of Political Science, which was meeting in 1976 in Edinburgh, the complex and timely central theme: "Space, time and politics." In preparing for that congress the political scientists decided to call on geographers, for they recognized the contribution which political geography had made and could make to that theme. At first, in 1975-1976, in examining the spatial factors in politics, attention focused on the evolution of the concept of territory and on spatial partitioning by boundaries and frontiers of all sorts; these were traditional concerns of geography and politics. Thus a notable rapprochement came about between geography and political science. It was formalized when, as an outcome of its Edinburgh Congress, the International Political Science Association established a committee to study political geography. The first Round Table of this committee gathered in Paris to discuss "centre and periphery." About twenty scholars took part, coming either from the fields of geography or political science, or even historical studies or sociology.

Historians had long ago recognized the need to ground their analysis on a solid geographical substratum; now, the evolution of most social sciences requires some accounting of the spatial variations encountered within society. People located in different places understood differently, debated and applied in varying manner what seemed to be the same ideas, theories, and policies. As the study of politics has grown more international, it has become concerned with the reasons for and the significance of spatial variation in political behavior and the use of geographical factors in political thought and analysis. Since the beginning of this century these questions have permeated the study of electoral patterns. As the twentieth century grew older, well-established imperial and national structures were shattered or crumbled. New political hierarchies have arisen. With rejuvenated vitality, old regionalisms have challenged national centres. The relationships between centres and peripheries have thus become a major source of tension, conflict, or at least political debate. In the developing turmoil, political scientists are asking new questions, seeking direction for some stability and order in the theory and practice of politics.

Contemporary geography has shown an increasing interest in the political factors of geographical phenomena. As it has evolved from descriptive and analytical attitudes to more prophylactic concerns in the study of mankind and its environment, geography has increasingly recognized a long-standing and permanent relationship with the political process. The same space that constitutes the geographic environment coincides exactly with the space that is the concern of the political process. And it is through legislation and politics that most of the partitioning, organization, and expansion of the geographical space are achieved.

The time seemed ripe indeed for the two disciplines to work together. The concept of centre and periphery, which implies both a specific relation of complementarity and some possibility of opposition and confrontation, proved to be useful in exchanging and integrating the knowledge and approaches of geographers and political scientists. I shall not attempt a summary of the matter in this preface. The reader will be well informed in the rest of the book. However, we did not intend to cover here all the aspects that a full examination of the centre/periphery relationships may involve. We hope to have provided in this book an instrument for further study and analysis of a rather exciting concept.

As editor, I wish to express my deep gratitude to all those who have contributed to this book—and they are many. First, I wish to express my sincere appreciation to the scholars who have contributed manuscripts. The twelve authors of this book come from eight different countries and have often revised or even rewritten their original papers for this publication.

They deserve to be commended for their cooperation and patience. Similar thanks are due to another dozen participants of the Paris meeting who contributed to the discussion, and who in some cases wrote special papers and have shared, to some degree, in the final version of this book.

Our sincere thanks are due to the Council of the International Political Science Association, the Fondation Nationale des Sciences Politiques in Paris, and the Maison des Sciences de l'Homme of Paris for their sustained interest and generous support, which have made possible the work of the committee, the meetings in Edinburgh and Paris, and the publication of this book. I must express special gratitude to my colleagues on this committee, who were also convenors of the Paris meeting: Professor Jean Laponce, University of British Columbia, and Professor Stein Rokkan, University of Bergen, both past-Presidents of the International Political Science Association. We mourn Stein Rokkan's untimely death in 1979.

Our special thanks go to Miss Margaret Loveless for redrawing some of the maps, and, last but not least, to Miss Sheila O'Clarey of the School of Geography in the University of Oxford for her tireless assistance in typing large parts of the manuscripts and maintaining the contacts with all those concerned with this book at its various stages.

—J.G.

# 1

# CONFRONTING CENTRE AND PERIPHERY

Jean Gottmann

Geography and politics have long been in search of each other. Scholars interested in the study of politics and geography met to discuss the concept of centre and periphery and the problems they saw arising from this relationship. They found the experience fruitful. The debate seemed to contribute an improvement to our ways and means of assessing and analyzing the political processes around us.

That "centre and periphery," a model originating in geometry, could have been used so successfully to bring together geography and political science is a sign of the times; it is significant for the understanding of the modern evolution of thought in the social sciences. Besides a concern for mathematical tools to reason with, the social sciences demonstrate increasing interest in spatial factors. Phenomena observed either as political or geographical appear to have four permanent characteristics in common: (1) they develop in the same space, that is, in the space available to human activities; (2) they reckon with the diversity of that space; (3) they develop according to decisions influenced by the perception of spatial circumstances; (4) they reckon with a considerable diversity of the ways in which these circumstances and what should be done about them are perceived by people.

Geographical factors, in terms of spatial variation and the means of manipulating this variation, have been recognized in politics as far back as records go. They are repeatedly mentioned in Herodotus, Plato, and Aristotle. They may have receded into the background during the age of spirituality, when political behaviour claimed to be directed by faith alone, but they came back into the limelight with the Enlightenment. Machiavelli recommended that the Prince spend some time hunting in order to keep a feeling for the *terrain;* and he paid attention to the spatial distribution of the population.

Political factors were basic to geography from its first steps in studying human phenomena, insofar as these were conditioned by rulers, laws, frontiers, and regulations. "Region" came from the same Latin root as regulation, regent, and regal. Politics and police were both derived from the Greek *polis*—the city. Obviously political authority organized the use of geographical space; and the organization of space evolved through a political process. The political institutions exercising authority had to be located somewhere. That authority operated within certain spatial limits, even though certain sovereignties may have at times claimed either the whole globe or a few vehicles moving in interplanetary space.

Location has thus played an important part in both geography and politics. The study of geography has largely been concerned with local circumstances and, particularly, with a precise definition of location. Indeed, how could spatial variation be observed, represented, and assessed, and spatial relationships established unless every point or area could be clearly located within the whole space, considered in such a way that it could not be mistaken for another. The need to define every location in a way establishing its uniqueness and its spatial relations with other points is especially indispensable to navigation, to movement between specified places. The position of travellers could first be established with relation to well-known and easily recognizable landmarks. As mariners began to strike across undifferentiated marine space, out of sight from the shores, the need developed for systems of mathematically calculated coordinates.

Geometry supplied the systems of coordinates, helped by astronomy in the task of measuring our planet. Much of ancient science in the fields of mathematics and astronomy was worked out to satisfy the needs of navigation for trade, migration, and the search for more resources. More mathematical skills emerged to answer other needs of trade, such as measurement of weight, volume, and distance. As trade developed, political authority followed it to protect the traders and regulate the transactions.

This historical process continues, interweaving human geography, politics, and the systems of coordinates determining location. The concept of location, however, is growing more and more complex and increasingly multifaceted, because the purposes are more and more varied for which location can be used, and that use or the access regulated. The purely geographical system of coordinates, using latitude and longitude to define the position of any point or system of points on the earth's surface, may still fit navigational requirements (with the addition of altitude and depth measurements for air and sea navigation), but the political concerns about geographical space and location expand in variety and complexity as a result of many trends. Among these trends are the consequences of modern technological, economic, and social change. They cause instability and constant interference by political authority into new fields of human activity. The instability prompts the agents of authority to maintain or modify situations perceived as orderly. For this category of problems, the relationships between centre and periphery have come to provide new systems of coordinates.

## THE GENESIS

The need for a debate of the centre and periphery model in political geography may now appear as a rather logical response to the demand generated by contemporary political evolution. Several more specific trends were, however, at work to produce this outcome besides a few more random circumstances. The search for new systems of coordinates and the remarkable current fashion of the centre/periphery model may be motivated by a series of concerns, some of which are worth reviewing here.

First, the social sciences have now for some 30 years been experimenting with abstract, theoretical, and, if possible, mathematical methods. By the middle of this century, a person of scholarly inclination could not help being impressed with the immense practical consequences that had emerged from scientific research in the physical and bio!ogical sciences using mathematical and statistical methods of analysis. The social sciences were bound to attempt endeavours in similar directions, and after 1950 they did so increasingly. The vogue of quantitative and more or less mathematical methods, leading insofar as seemed possible to "experimentation," came to geography and political science only in the 1960s, but developed fast.

Second, both of these disciplines have been using statistical analysis in certain areas at least since the nineteenth century. The geographers, working on the problems of population pressure, had early developed and widely applied the simple mathematical model of the density of population; this index, combining area and numbers of inhabitants, became a useful and popular tool of measuring certain aspects of spatial variation. Excessive assertions as to its political and economic significance led, however, to discredit, especially when high population density was used to bolster claims of territorial expansion as was done by expansionist powers in the

1930s and in World War II. Theoretical attempts at establishing an "optimum" value for population density were also unsuccessful and left the impression of a systematic exaggeration of the practical value of population density indices.[1]

In the same period, political geography initiated another aspect of cartographic analysis of statistical data of considerable political portent: the beginning of geographical analysis of voting patterns in the early twentieth century. Electoral geography first developed in systematic fashion in France, under the direction of André Siegfried.[2] It has since been generally adopted. With the help of computers, this statistical and geographic analysis of electoral behaviour has become a tool of great strategic value for both political science and practical politics. Relatively stable variations in voting patterns according to area were often observed and analyzed.

Third, by 1950 the social sciences felt a strong urge to reorganize their methods and purposes along lines that had yielded such spectacular results in the more exact sciences. Statistical and quantitative methods have been very useful, especially in the biological and medical spheres. But in physics it was theory based on formal logic and tested by experimentation that led to the main breakthroughs. To advance their own research, the social scientists attempted new tactics. More systematic, formal analysis came into vogue and helped new fields to develop or old sectors of research to advance in various directions. Experimentation in the political area was, of course, rather difficult without launching into *Realpolitik;* the same was true for human geography, especially in regional or urban matters. However, there was a possibility of simulating certain situations for complex human aggregates, at least on paper, where abundant quantitative or measurable data were available. Regional and urban planning has been doing precisely that for some time, before launching experiments in vivo with new settlements. Now simulation could start with the simplest kind of situations, eliminating the excessive multiplicity of factors at play in the real world. Both geographers and political scientists have indulged in simulation and planning games on a large scale in the last 25 years. In that endeavour they have widened the scope of their studies, produced much for the practitioners to digest, engaged in predicting, and have somewhat modified the whole framework of their disciplines.

Fourth, this modification has only begun. These are dynamic times and the momentum of evolution does not show signs of abating in terms of the nature and range of the political problems to be solved. These problems are bound to attract the attention of scholars and provide direction to their endeavours. However, the great movement of reformation taking place in the social sciences was not determined by the sheer pleasure of emulating the exact and natural sciences. The idea was and remains to progress in the elaboration of theories and formulas that will help solve the problems of society and the riddles facing the politicians.

Some problems are indeed permanent and will continue to require the use of old and renovated means of investigation: e.g., electoral behaviour, which will maintain a general interest in the analysis of voting patterns; or the distribution of population with regard to available resources, a complex relationship that could not be handled any longer by such simple though significant measurements as population density, migration trends, and trends in employment. In our era of emphasis on ethics and equality, more attention has to be given to regional distribution or spatial variation. Here the notions of centre and periphery came into the fore, owing to political rather than geometric reasons. The simulation of experiments in modifying existing economic structures in a variable space became useful to prepare planning policy.

Fifth, any political and economic structure at work necessarily involves elements of centrality. How is the centre to be defined, however? In a political spatial unit such as a state, a province, or a county, the centre is normally the place where the seat of authority is located. It is usually called the capital. In a rural area it may be the courthouse wherefrom justice is administered. In a city it will be the city hall and such institutions as the courthouse, market, and bank, which usually locate rather close to one another. Their grouping will make that area central in the city because the citizenry will need to come there often to perform transactions. If a large number of citizens have no need for the kind of transactions that may be best performed in the central district, they will still recognize it as the city's centre because decisions affecting them are made there, because many roads and means of transport converge there, and because of the institutional and historical status of that area.

Centrality must be perceived by the people using the "central place" as such. That perception is not the simple result of physical design apparent in the field; it is influenced by historical knowledge, political organization, economic functions, and so on. Within a national territory there could be several important centres well recognized by the population as national centres: one could be the political capital, another the major economic and financial centre, still another the seat of the higher judicial instance, and perhaps still another the locus of specialized political and cultural activities that complement those of the other three centres. In the case of Switzerland, a small country with a pluralistic confederal structure, Bern, Zurich, Lausanne, and Geneva seem to correspond to the four different central functions suggested. Many nations, regions, and even cities have adapted to a more or less pluralistic network of centres.

Sixth, the peripheral position seems easier to recognize because it is more diffuse, more widely distributed. It would seem that any area or location depending on a centre outside it for services or decisions affecting the life of its population feels itself to be on the periphery. In geographical terms the periphery is what surrounds the centre, a geometrical relationship; the farther away a point is from the centre, the more peripheral it would be. But the political relationship is different: peripheral location means subordination to the centre. In a stable, orderly situation the subordination is accepted even if somewhat deplored and resented. A lack of resignation to such subordination would obviously lead to conflict and instability.

The instability may result from political ideas, a shift in the psychological attitudes toward an established order. It may, however, be triggered by rapid technological and economic change, modifying the spatial distribution of activities, population, and resources. Studying central places in southern Germany, geographer Walther Christaller offered a method of organizing space and analyzing an established spatial pattern of centrality functions through a set of central places endowed with a clear hierarchy, based on distance from the largest city and on the size of the smaller centres. This German school of thought could be traced back to the pioneering work of Von Thuenen in his book on *The Isolated State,* which started from the assumption of simplified centrality: "Imagine a very large town at the centre of a fertile plain, which is crossed by no navigable river or canal." This was the simulation of monopolistic centrality amidst uniform conditions of transport. The enfolding economic analysis could easily show that the farther away from the centre a location would be, the more economically disadvantaged it became.[3]

But here again the political factor would immediately introduce different variations. The earliest and most consistent themes of study in political geography have been capital cities on the one hand, frontiers and boundaries on the other. It was a politically qualified centre and periphery relationship. However, for reasons of national defence and international trade, the frontier regions often acquired a special role in the national structure, and while peripheral by definition, came occasionally to dominate the whole system, of which they were a crucial periphery.

The brief chapter by Owen Lattimore recalls two such major evolutions in the past, in the ancient Chinese and Roman empires. Many such precedents are known in history. Politics makes for a constant instability of the centre and periphery relationship. The purely geometric concept may be misleading, even though it provides a stimulating imagery to start with.

# THE CHALLENGE TO CENTRALITY

In political geography the centre and periphery image conveys two rather different ideas: first, a symbol of systematic organization of space around the notion and through the function of centrality; second, within that established order, the opposition between the dominant centre and a subordinated periphery, suggesting the possibility of confrontation. In the political climate of our century, the premium put on economic and social equality seems to call for a restructuring of such spatial distribution as hinted at by the opposition of centre and periphery, favouring the periphery.

The fashion of the centre and periphery model started with scholars because of the geometric quality of the relationship, the simplicity, and the clarity of the design. However, it spread largely because the connotations of inequality, unfair established order, and potential confrontation appealed to the modern impulse of formulating a clear-cut problem in order for a situation to be righted. Whether centrality could be eliminated or spatially spread out, diffused to all the periphery—at least to the inhabited periphery—is a difficult question to answer. Some of the essays that follow in this book discuss the matter and hint at the position of the question in specific cases. Centrality is certainly being challenged. There are several reasons, both geographical and political, that may have nowadays intensified the challenge.

First, challenges to the established spatial order result from the dynamic trends of change characteristic of recent decades: technological evolution, economic growth, the redistribution of income, a shifting occupational structure, welfare reforms, and new ideologies all combined to produce a feeling of relative liberation from old shackles. The quick pace of multifaceted change opened the perspective of a fluid, malleable world environment that could be moulded to obtain a new order with greater justice and more equality amongst people or between regions, if the material opportunity was properly manipulated. Peripheries, for instance, could acquire new centrality. All existing knowledge should be abstracted and reclassified to achieve either homogenized equality or new, more desirable, hierarchies. What was considered utopia yesterday was now described by many as realistic. Centre and periphery were convenient notions to use in formulating policy, outlining plans, and reordering theoretically spatial and political patterns.

Second, within most countries, rapid technological and economic change led to massive urbanization. Active redistribution of population, economic activities, and political weights ensued. Urbanization increased concentration around the major urban centres, thinning out large areas, deepening the imbalance between regions, stressing contrasts, and arousing more opposition between centres and peripheries. The same process of urban growth expanded many cities to such a scale that, as suburban rings developed, the urbanized districts became metropolitan areas within which differences sharpened between central nuclei and peripheral sections. Whether on the regional or national scale, modern urbanization considerably reshuffled locational values and the previously accepted geographical order. Hence the urge to reconsider relationships of the centre/periphery kind.

Third, technological and economic evolution set many robots to expedite tasks that used to be performed by hard, constraining, physical human labour. But in exchange it unleashed on mankind an immense flood of information, made of myriad bits of information of all sorts, diffused easily through space by modern mass media and other communication networks. The effects of this "information revolution" compounded those of the belief in the "malleability" of the environment. These trends concur in calling for more generalized, abstract methods and concepts to help classify, interpret, and process all the available data. To process information relating to spatial variation and location, the centre and periphery model appeared to be a valuable instrument.

Fourth, while all these "permanent revolutions" were going on and more modification in the perception of what was possible and fair occurred as a result of the recent concerns about pollution, resources exhaustion, limits to growth, ecological preservation, and so forth, the major impact on political attitudes came from economic factors and socioeconomic theories. The political and strategic concerns that used to dominate the thinking of governments and the reactions of peoples receded into the background in a long period of relative peace between major powers. Although local and regional wars have occurred, the last third of the twentieth century seemed on the whole relieved from the fear, almost constant in the past, of large-scale military confrontation between great powers.

This relatively peaceful period of the last 30 years has seen the more frequent sources of trouble in the tensions and conflicts of a political and economic nature develop inside national structures, acquiring a regional character, even though occasionally the conflicts transgressed international boundaries. In such situations the centre/periphery dichotomy proved more handy than the older concerns about frontiers and capitals.

Fifth, all the forces working for change remapped the political systems around the world. More new states have risen to independence in the last 40 years than at any time in history; more new federal or regional structures have been formed, whether within nations or on a transnational level, than for centuries in the past. Each of these new political entities established a capital and sometimes conferred central status on several cities. The number and functions of the seats of governmental authority have been greatly increased in this process. New relationships of the centre/periphery type have proliferated around the world, making political geography more dynamic and fluid.[4] Whether the creation of so many new centres reinforces or weakens the very phenomenon of centrality remains unclear at this point but becomes more interesting to discuss.

At least four chapters in this volume touch upon the debate arising from the proliferation of political centres: Lewis Alexander outlines the search for centrality in small island systems that are really in search of their proper personality; George Hoffman describes the centre/periphery struggle in the centrally planned regimes of the socialist countries in Southeast Europe; Nirmal Bose hints at the need for organized centrality in Southern Asia, emancipated from the British imperial system; Francesco Compagna and Calogero Muscara show the use of regionalistic claims and periphery versus centre concepts in the Italian political game and the impact it may have on the unity of a nation with a long history of political division. These are four very different cases. They concur to illustrate the proper characteristic of our time: it seems that, for a variety of reasons (ethnic, purely historically cultural, recently economic, or essentially political), various peripheries or subordinated centres rebel against established hierarchies, i.e., previously accepted dominance either by central places or especially powerful groups. Usually these are geographically identified with the place where the agencies are located through which dominance is exercised.

If one applies formal reasoning, it appears unrealistic to assume the possibility of spreading centrality in such a way that it would be distributed equally everywhere. To suggest that all formerly central and peripheral places could become equally central seems a contradiction in terms: such an attainment would eliminate any need for this kind of concept as either centre or periphery to achieve a homogeneously diffused state. Mankind could hardly subsist in such conditions. Equality among people does not mean full uniformity; it is an aim that can be achieved to a relative degree, relative to a purpose of human activity.

The challenge to centrality expresses, however, a deep psychological aspiration of our time, and perhaps of all times, but which was only recently able to emerge with its present great strength. The need of centrality to organize the use of space remains in practice as strong as ever, though more complex because more pluralistic in nature. It is one of the most interesting conclusions of Raimondo Strassoldo's chapter, reviewing the many uses and meanings of centre and periphery, that perhaps mankind will have to learn to live with a multiplicity of centres. Stein Rokkan's very personal review of the history of Western Europe already assumes *de facto* that Europeans, and others, have long lived under the influence of more than one centre in most areas. About 30 years ago, proposing to postwar human geography a new method of analysis, I wrote about the need to visualize space as organized around chains of crossroads.[5] This sort of vision is increasingly appreciated today, as noted in most chapters of this book that look toward a pluralistic system of centrality.

## MANAGING CENTRE AND PERIPHERY

Perhaps the most accurate way of summarizing the approach of political science to the centre and periphery model is indeed to stress a confrontation, but one that ought not to lead to dislocation. In geometry, and just as much in geography, centre and periphery complement one another: there is no periphery unless the spatial figure considered has a centre, or central sector; inversely, once a centre is determined, there is to be a periphery around it; otherwise of what is it the centre? This essential interdependence does not prevent, of course, the confrontation once politics is introduced into the spatial pattern.

In Chapter 2 Raimondo Strassoldo brilliantly reviews the many different meanings and uses of the model. He dwells with caution and great erudition on the Western beliefs about centrality, a notion associated with the sacred. He stresses how perspective in Western art led to a glorification of the centre or focus. Oriental civilizations seemed to accept an almost unlimited number of centres. His conclusions are interesting and somewhat poetical, but pessimistic. Strassoldo's interpretation of the organization of space in the past may be too formal and too idealized; in practice, even the Western world has always lived with a plurality of centres. Still, the modern stress of the periphery as the hero is something unique.

In Chapter 3 Paul Claval in terse and metaphorical style, takes up critically the use of the centre/periphery model. The uses with which he is concerned are more specifically those found in a few contemporary schools of geography and political science. To the philosophic approach of Strassoldo the geographer Claval opposes a more demanding methodological approach, and he seems to conclude that the model has been of only limited help as yet in solving the problems of political geography.

But is it the role of theoretical models to solve problems? Or should they simply help to formulate and clarify the problem? Alan Henrikson illustrates the latter proposition within an enormous example: the United States was on the periphery of the world by 1900; by 1950 it was at the centre of the world;

and where in this relation does it stand now? A careful diplomatic historian by training, now specializing in the political perception of spatial symbols, Henrikson reviews America's debate on its own role in this century, quoting the use of the centre/periphery opposition even by President Carter. His conclusions do not answer the question clearly, as he too wonders whether the world is becoming polycentric. However, Chapter 4 shows well the usefulness of the model in putting the problem in perspective.

A very different inquiry follows in Chapter 5, in which Francesco Compagna and Calogero Muscara analyse the centre and periphery debate in Italian politics as it affects the implementation of regionalistic policies within a recently united and still very diversified nation. In some respects, their specific case study illustrates some of the points made by Strassoldo and Claval from more theoretical approaches. Compagna's concern is at times that of a practitioner, as this professor of geography has also been one of the spokesmen of the southern "periphery" and a specialist on the regional reform in the Italian Parliament. This chapter demonstrates how, in a frail political framework, peripheral claims may come into conflict with sheer patriotism. It is also somewhat paradoxical to see extreme regionalistic and decentralizing views held by the Marxist left which, once in power, could not fail to impose strongly centralized planning.

The impact of Marxist trends propounding the periphery versus centre philosophy was mentioned also by Strassoldo and Claval on the level of general ideas. In Chapter 6 George Hoffman discusses the peripheral claims and regional planning precisely in the communist countries of southeast Europe. Each of these has recently been a new and dynamic experiment in the setting up and readaptation of centre/periphery relationships. The spatial variations appear deeply rooted in the cultural and political history of each region. The countries of southeast Europe had long been peripheral within empires centered in Constantinople, Vienna, or earlier, Venice. They have had to create their own centres of a modern type, away from their traditional mountainous heartlands.

Another search for centrality in new patterns of territorial organization, and an apparently more elusive one, is provided by the need for political capitals and other forms of centrality in groupings of islands that have recently acceded to self-government. Lewis Alexander is a political geographer who has specialized in the legal and administrative aspects of the organization of maritime, including insular, space. His Chapter 7 offers a good example of the striving for centrality and of the dilemmas for people and areas threatened with a status of periphery in the organization of small, isolated island systems. The conclusion raises once again the question of the importance of networks of places endowed with different kinds of central functions. Many of these networks appear commanded from centres outside the island system, which needs to participate in these centres to link itself to the outside world. The dilemmas spread and multiply.

The difficulties of adapting centrality to the needs and claims of the peripheral regions, which "orbit" around the centre and wish to benefit from it without damage to their own interests, take on many different aspects. In many cases these difficulties arise within the structure of the central place itself, especially when it is a large and growing city. Jean Laponce, who is deeply concerned with the entanglements and tensions arising within the centre/periphery modelling, offers in Chapter 8 a skillful and perceptive analysis of the concept of centrality: first he tells us of a game analysing the ways in which centrality may be perceived in a group of individuals; he then questions the perception of the centrality of a city such as Montreal by newcomers settling there after migrating from much smaller, peripheral places. The questioning of the perception of the centre, envisaged in rather theoretical fashion in this chapter, provides a respite to rethink the human impulse toward centrality after the series of specific cases of what may be called the "political management of centrality" examined in Chapters 4 through 7. Jean Laponce brings into the question of spatial organization the complexities often arising in large cities from the diversity of languages and cultures.

But managing centrality is an exercise in government administration, and in a situation in which a spirit of confrontation exists, it will be considered an exercise in domination. The attitude of confrontation is rather widespread and loudly advocated in our time. It may not have been the first such time in European history. To see the role of networks of central cities in Western Europe and to attempt a classification of peripheral situations, Stein Rokkan reviews in Chapter 9 the political and cultural divisions of Europe since early medieval times. The role of the ethnic and linguistic puzzle imprinted on Western European lands by many centuries of migration, dynastic feuds, and unequal distribution of resources brings out the need for flexibility and diversification in the centre/periphery political relationships, lest constant conflicts and tensions tear the countries apart. The mass of data available on the thousand years of European history involved is immense. In a brief essay, the choice of data used or emphasized is perforce arbitrary. Many descriptions may appear highly subjective, but this Scandinavian view of Europe's past is challenging and deserves to open a debate of its own.

All chapters testify to the essential role of historical data in understanding the centre and periphery relationships and in illustrating the situations studied. The more stubborn peripheral opposition to extant political structures, and therefore to current dominance of certain centres, is found among peoples who can claim a long past of separateness, differentiation, and/or oppression. Seldom does a periphery rise, for new reasons of discontent, that had not considered itself for sometime culturally different from the dominant centre.

The power of the historical factor is also illustrated in Chapter 10 by Owen Lattimore. This great specialist of Chinese and Central Asiatic affairs, past and present, chose his main case in the northwestern reaches of China, a lifelong, preferred subject of his research. The proposition Lattimore demonstrates is the frequent role of the periphery as an innovator, modifying existing territorial structures in politics. The example of the Chinese Empire, so often dominated by peripheral elements, is probably the most striking throughout a long history; but the evolution of the Roman Empire is another such case, and in modern times the rapid emergence of the United States, coming from a peripheral position, is again an illustration of the same trend, as demonstrated by Henrikson in Chapter 4. Indeed the use of peripheries against the centres that dominate them politically, being an innovation on the existing map, mainly causes, if the lessons of history are to be repeated, shifts of dominance to new centres located in previously subordinated areas.[6]

History admits, however, a whole gamut of evolutionary solutions to problems of confrontation. The growing political division of Southern Asia, since that region's emancipation from the large-scale centralized system established by British colonization, is studied by Nirmal Bose, from an Indian viewpoint, in Chapter 11. Except for Henrikson's discussion of a worldwide aspect of centrality, none of the other chapters examines a case on as large a scale as Bose's. Precision and specificity are all the more difficult as the number of variables and factors increases with the scale in human affairs; but the situation Bose outlines appears, in his opinion, to call for some federal transnational scheme to avoid more confrontation and further disruption of common interests, organizing in a vast spatial system neighbours such as India, Pakistan, Bangladesh, and Sri Lanka. The common denominator here seems to be proximity with a certain common historical inheritance left over by former empires in a part of the world of great human diversity.

Centrality may be challenged but it seems to manage partly by proliferating, partly by showing increasing concern for peripheral complaints, and mainly because it seems indispensable to the conduct of politics. Before concluding, we must now proceed to the different approaches and cases offered in the chapters that follow. If centre and periphery form a relationship that originated in geometry, it may now be that political geography and political science prefer topological approaches. Perhaps a new philosophy of political geography is emerging. Centre and periphery has been a stimulating model to apply to so many different concerns.

## NOTES

1. Population density was used in the late nineteenth century to describe crowding, or the lack of manpower, and the spatial variation of the pressure on local resources and land use in various countries. "Overpopulation" became a popular notion between the two world wars. At the same time the search for an "optimum" level of density of population attempted an "objective" index of what would be under- or over-populated areas. Already before 1939 a few geographers, such as Albert Demangeon (of the Sorbonne), showed that overpopulation was mainly a way of perceiving or assessing the need for more territory to develop more resources. The powers that claimed to suffer from overpopulation in the 1930s, especially Germany, Italy, and Japan, have enjoyed higher standards of prosperity after 1950, despite a much higher density of population. The opposition between over- and underpopulated countries was replaced after World War II by the distinction between well-developed and underdeveloped areas.

2. The pioneering work was Siegfried's, *Tableau des Partis Politiques dans la France de l'Ouest* (1915), a landmark in both electoral analysis and political geography. It was followed by a whole school of electoral geography and electoral sociology. In France the work of François Goguel has been especially significant. Now these methods are widely used in all democratic countries. For a recent brief survey, see M. Busteed (1975).

3. Von Thuenen's *The Isolated State* was first published in 1826 in Hamburg under the title *Der Isolierte Staat in Beziehung auf Landwirtschaft und Nationaloekonomie*. Its English translation has appeared as *Von Thuenen's Isolated State* (1966). Although it has seldom been mentioned, Von Thuenen had certainly been influenced by the work of the famous German political philosopher Johann Gottlieb Fichte, *Der Geschlossene Handelsstaat, ein Philosophischer Entwurf als Anhang zur Rechtslehre und Probe einer Kunftigzuliefernden Politik* (1800), an advocacy of the centralized closed commercial state, possibly inspired by Plato and certainly a forerunner of German policies endeavouring self-sufficiency and autarky. Christaller's *Die Zentralen Orten in Sueddentschland,* first published in 1933, and in English translation as *Central Places in Southern Germany* (1966) is, of course, a very different, less political, approach to centrality. However, its stress of regional hierarchy in a closed frame stems from the idea of the distribution within a closed space, advocated by Von Thuenen as analytical economics and by Fichte as political philosophy.

4. See Gottmann (1973); especially Chapter V, "Crossroads and Frontiers amid Modern Fluidity."

5. This was proposed in my article "De la Méthode d'Analyse en Géographie Humaine." In the same article I also suggested that geographers may sometimes arrive at representing a region by a mathematical model (*un être mathématique),* an ambitious prediction at the time. In 1948 I began teaching at the Institut d'Etudes Politiques of the University of Paris. Jean Laponce was then a student at that institute. He presided over the International Political Science Association when the Study Group on Political Geography began to work in 1974–1976.

6. It may be worth mentioning that reviewing the book by Jules Sion on *Les Paysans de la Normandie Orientale,* the French geographer, Paul Vidal de la Blache, emphasized (in the *Annales de Géographie,* (1909) that new towns arise at the contact of different but contiguous regions—a remark that may be generalized to teach that peripheral or frontier zones breed new centres.

## REFERENCES

BUSTEED, M. (1975) Geography and Voting Behaviour. Oxford.

CHRISTALLER, W. (1966) Central Places in Southern Germany. Englewood Cliffs, NJ: Prentice-Hall.

GOTTMANN, J. (1973) The Significance of Territory. Charlottesville: University of Virginia Press.

________(1947) "De la méthode d'ánalyse en géographie humaine." Annales de Géographie 301: 1–12.

HALL, P. [ed.] (1966) Von Thuenen's Isolated State. Oxford.

SIEGFRIED, A. (1913) Tableau Politique de la France de l'Ouest sous la Troisieme Republique. Paris.

# 2

## CENTRE-PERIPHERY AND SYSTEM-BOUNDARY: CULTUROLOGICAL PERSPECTIVES

Raimondo Strassoldo

### INTRODUCTION

Centre-periphery is a geometric, static concept. It can be brought to life for use in the social sciences in two ways:

(1) analysing the *symbolic meaning* of centre, periphery, and their encompassing notion, the circle; symbol-analysis can also be called semiology or, more widely, *culturology,* according L. K. Whyte's not too fortunate proposal
(2) using it as a pattern-system model of an *action-system,* in Kuhn's (1974) terms, the centre then becomes a living nucleus, a decision-making node, a control structure; in this case, it is "system," not "circle," that becomes the encompassing notion, and the whole panoply of systems analysis can be brought to bear

I have already attempted, in several places (Strassoldo, 1970, 1975, 1977, forthcoming; Strassoldo with Gubert, 1973), the second kind of analysis, placing the centre-periphery polarity in the context of a "general theory of boundaries" fashioned, after the manners of the General Theory of Systems, out of the insights and fragments of a wide array of established disciplines.[1]

Here I would like to develop the former approach, and give only synthetic summaries of the conclusions reached on the second level. For purposes of brevity, terms and arguments are employed without excessive concern for formal definition and articulation. The reader's kind insight, rather than logic, is called to task. More formal presentations can be found elsewhere. I think it not improper for the only sociologist in a symposium of political geographers and other serious scientists to assume a dionysian, rather than apollonian, posture.

## THE MEANINGS OF CIRCLES AND CENTRES

One of the most archaic ideograms is the neolithic Sun Wheel. The circle is the symbol of the sun, of life, of enlightenment, of the self, of the "totality of life," of the union of opposites, of human and universal perfection. The *mandala* is one of its more universal and sophisticated expressions. They can be found also in Europe, in the rose windows of medieval cathedrals, in the haloes of saints, and in the ground plans of temples and cities (Jaffè, 1976).

Mandala "means circle: the translations from the Tibetan sometimes render it by 'centre' and sometimes by 'that which surrounds' " (Elaide, 1969: 52). I think we have here, in a nutshell, most of issues and ambiguities that concern us in this essay. The mandala is a class of visual archetypes of the most subtle and elaborated cultural meanings; they express cosmological myths produced along thousands of years in a wide culture area. They are basically patterns of concentric circles and other figures, variously complicated in size, color, textures, and additional elements. They are circles, and therefore can be conceived as points, as (enlarged) centres; in addition, they *have* centres, but they also have a periphery, a boundary line marking them off the environment, the field.

All archaic cultures, Eliade assures us, have been fascinated by the symbolism of the centre, to which that of the circle is closely associated (the circle is generated by a centre). Art historians are most sensitive to this tradition in Western civilization, especially now that it has been destroyed in the arts. Centre and circle mean symmetry, proportion, and perspectives. Hans Sedlemayr has decried the "loss of the center" in modern Western visual arts, while Paulet (1961) has traced the "metamorphoses of the circle" in Western culture, beginning with Plato: "The notion of a circular pattern, mirroring the parallel organization of the Cosmos, structured in concentric circles, seems to be the most typical element of the Platonic tradition." It was elaborated on by the neo-Platonic school of Plotinus, Proclus, et al., who spoke of the world as a circle, and of the centre as its "cause," i.e., God. It was taken over by the early Christian philosophers, who called God the Centre, and revived in the Renaissance by Marsilio Ficino ("the Soul is the

Centre of Nature"), Pico della Mirandola ("God has placed Man in the Centre of the World"), Paracelsus ("all universe surrounds Man, as the Circle surrounds the Centre"), and Giordano Bruno ("the Soul is a sort of Circle").[2]

## THE DIALECTICS OF CENTRE AND BOUNDARY

Drawing from his studies of non-Western cultural systems, Eliade (1969: 54) makes two basic points. The first concerns the dialectics of centre and boundary (periphery). On the one hand,

> every human being tends, even unconsciously, toward the Centre, and towards his own Centre, where he can find integral reality-sacredness. This desire, so deeply rooted in man, to find himself at the very heart of the real—at the Centre of the World, the place of communication with heaven—explains the ubiquitous use of the "centres of the world."

It can be noted, in passing, that a similar view about the function of centres is presented by Ardrey (1966) in the most fascinating final chapter of his controversial book, *The Territorial Imperative*. Building upon a metabiological theory of animal needs and the functions of territory, he suggests that while the frontier provides opportunities for conflict, adventure, social encounters, discharge of aggressiveness, and self-realization (stimulation and identity), the centre provides security, rest, and the environment favourable to reproduction of the species (mating and rearing of the offspring). And insofar as the security of the individual and the continuation of the species are the ultimate values in nature, the centre can acquire a sort of biological sacredness.

On the other hand, *groups* build boundaries between them and the outside. Circle and centre are the sacred loci of identity and security of every human being; but human beings are social creatures, and their individual spheres coalesce into an ordered cosmos: "At the limits of this closed world begins the domain of the unknown and the formless. On this side there is ordered space; on the other, outside this familiar space, there is the unknown and dangerous region of the demons, the ghosts, the dead and the foreigners—in a word, chaos or death or night" (Eliade, 1969: 37).

According to such *Weltanschauung*, which can be found in most great civilizations such as China, Mesopotamia, and Egypt, there is then a fundamental distinction—a *boundary*–between *cosmos,* the ordered, familiar world inhabited by fellow men and *chaos,* the frightening outer world of evil forces and monsters. Sociologists such as C. Cooley have called our attention to the basic, if rather bland, parallel distinction between "ingroup" and "outgroup"; and depth-psychologists have generally interpreted such fundamental and almost universal dychotomization of the world as projections of destructive drives into the environment, as construction of the external "enemy," and as an extraflexion of our own darker inner side. The outsider is a proper object of our contempt, hate, fury, and terror; he allows us to show our righteousness and virtue. The sharing of a common enemy becomes a powerful cement of social solidarity. G. Simmel and others have highlighted this function of social conflict and the social status of the stranger, and C. Schmitt has maintained that the category of the "enemy" lies at the foundation of political life. The boundary line between "us" and "they," between ingroup and outgroup, between countrymen and foreigners, between men and barbarians, between one's own world and the external environment, between inside and outside, appears then as one of the basic moral categories.[3] It is a boundary encircling the individual self as well as whole societies; every level of social organization is marked off from its environment—more or less hostile and frightening—by such a psychocultural boundary. The strongest one is usually found at the level of cultural area or civilization; almost as deep and consequential are the sacred frontiers separating the political systems in the age of nationalism.[4] As soon as boundaries of this sort are conceived or discovered, defensive reflexes arise; mere de facto differences become objects of intentional, organized activities of promotion and defense. "Ecological" or "informal" systems begin to differentiate into "leading" or "control" subsystems, i.e., *centres*. So the defence (and expansion) of boundaries creates the conditions for the development of centres.[5] This is a common enough historical experience; assertive cultures become military nations, besieged cultures develop strong social cohesion and a political-military organization.

The second of Eliade's basic points however is wholly culturological. Most cultures have a three and four-dimensional image of the world. The earth might be flat, but the cosmos is a sphere or a set of spheres, and the centre of the world is usually the point of intersection between heaven, earth, and hell. "The most widely distributed variant of the symbolism of the Centre is the Cosmic Tree, situated in the middle of the universe, and upholding the three worlds upon an axis" (Eliade, 1969: 44). This myth can be found in ancient India, China, and Germany. Elsewhere we find cosmic mountains, towers, poles, or mounds. One of the most familiar and elaborated of such three-dimensional concentric cosmologies is the medieval European one described, for instance, in Dante's works. The centre is not only three-dimensional; it is also often dynamic, so to speak. It is the growth centre of the universe, from which it was created. Rabbinical texts, matched by Indian ones, state that God created the cosmos starting from a "navel" (Eliade, 1969: 43). Theories of the "big bang" are not that new.

## IN SEARCH OF STRUCTURAL GROUNDS

What seems to differentiate Western civilization from all others is the insistence on the symbolism of the *unique* centre. One way to explain it is the well-known idea that Western thought is "little more than a commentary to Plato"; we have already mentioned the importance of this tradition. What remains to be analysed are its roots. They are to be found, of course, in the Pythagorean religion of numbers and forms, where the circle had a prominent place as the image of divine perfection, self-sufficiency, incorruptibility, harmony, and the like; but is there something less cultural beyond it? Just why should the circle carry such meanings and exercise such fascination on human beings? Are there biological, psychological, or sociological bases to it?

Emile Durkheim was fond of looking into social structures for the sources of mental categories, values, and images. According to his analytical stance, it is not to be excluded that Plato's and others' sacralization of the circle has to do with the physical shape of the Greek polis, as of most other settlements, and with his own aristocratic background that of course entailed a central position in public gatherings, in the public political doctrine, and possibly also in residential location. The circle would then be the formalization of urban structure and the planar projection of the pyramidal political hierarchy; the centre is the place where sacred institutions and powerful families are located.

This Durkheimian approach seems to stand in sharp contrast with the culturological one, according to which it is mental images such as those of centre and circle that lead to the construction of round cities with social centres in about the middle. Eliade (1949), for instance, is emphatic in his opinion that the shape of walls, ground plans, and city centres is primarily of symbolic nature and that their utilitarian and esthetic aspects are only accidental and derived; he is especially concerned about the magical meaning of the defences of settlements: "ditches, labyrinths, ramparts, etc. were set up to prevent the incursion of evil spirits rather than attacks by human beings" (Eliade, 1969: 39). His position is radicalized by Jaffé (1976: 272), who insists that "whether in classical or primitive foundation the mandala ground plan was never dictated by considerations of esthetics or economics. It was a transformation of the city into an ordered cosmos, a sacred place bound by its centre to the other world." By the same token, it is controversial whether it was pyramidal social structures that gave rise to the geometrical concept of pyramid, or vice versa. The Durkheimian approach seems validated, however, by the fact that there are sound and obvious physical and behavioral reasons to account for the circular shape of settlements and public gatherings. Urban geography and human ecology have developed convincing theories about concentric growth of cities, due to competition for the most accessible locations, and theories have been developed about central places: Doxiadis' Ekistics postulates a "Central part" at every settlement level. Of course, consideration of human values cannot be excluded from these analyses; it has been repeatedly demonstrated that one of the main motives of the competition for location in the central business district is not really accessibility or other such "objective" factors, but *prestige*—the lingering symbolic lure of the sacred centre. Overall, however, it does seem that genuine ecological forces tend to produce urban patterns with centres and peripheries; just as it seems that the operation of market economic systems (the human analogue of ecological system) tend to produce hierarchies of settlements around "centre," "poles," or "capitals," and that the shape of these territorial units tends to the hexagonal pattern, which of course is the result of the compacting of circles. Thus, at the most primitive level, it might be that the idea of circle is the result of the economic-ecological forces that suggested the circular shape as the most economical one in building fences and walls, because it minimizes the ratio between perimeter length and enclosed surface; and the idea of sacred centre may be a cultural elaboration of the fact that this location was the most secure from outside harassments and at the same time the most internally accessible. Descending a further step in the level of naturalistic explanation, we can take notice that centre-periphery polarities tend also to emerge among social animals, both for defensive reasons (the ring of elephant or bison bulls around the soft core of cows and calves) and more interestingly, for communication purposes. Members of baboon troops tend to stay *around* the dominant male in order to have a better and unobstructed view of his behaviour, from which their own depends; he tends to sit in the middle, often on an elevated position, in order better to survey his domain. Baboons that for some reason are marginal tend, not unexpectedly, to stick to peripheral locations.

Jolly (1972) has detected two categories of primate social groups; the centripetal and the a-centric ones, and has rephrased the whole of status, and much of the rest of social structure, in terms of *direction of attention*.

Pursuing further this line of biopsychological arguments, one may speculate, a la Desmond Morris, whether breasts and nipples have anything to do with human delight in circles and centres; the visual arts certainly suggest it, and Freud associated roundness with femininity. But the mind boggles at the thought of the links that extreme Freudians like Geza Roheim may have discovered between the Platonic celebration of those shapes and the Hellenic fondness of callipygian youths.

On quite another level we find hypotheses drawn from the physiology of man's senses of spatial orientation. As a visual animal, man relies very much on sight for his perception and conception of the world; and the eye happens to be equipped with a quite distinctive centre-periphery polarity. It is currently suggested that we really have two visual systems: the retino-tectal and the retino-geniculo-striated. The first is based on the rods, and gives us the global image of the field in a wide but relatively fuzzy way; the second, based on the cones, is the organ of *focus* by which we fine-scan objects placed in the centre of attention. The fovea, a central area in the retina filled with cones, is its seat. Our visual sense of space emerges from the combination of these two systems within each eye, as well as from binocular vision (Trevarthen, 1968). Our total sense of space requires, in addition to the eye, the inputs of other organs and, indeed, the whole body. Paillard (1974) and others have demonstrated the importance of the vestibular apparatus in supplying the fundamental vertical spatial axis in relation to the mouth-head-spine-anus polarity, while Piaget (1948) and others have insisted on the tactilo-cynesthetic system of spatial orientation, which supplies information on the dislocation of limbs and body parts in relation to a theoretical centre of the individual, but also on the distance between a point in space and the individual, as measured by the amplitude of the movements necessary to reach it.

## THE TRIUMPH OF THE CENTRE: THE WESTERN SCIENCE OF PERSPECTIVE

Thus the full sense of space, as architects have long known, implies the time dimension and movement. But painters can work only with the two dimensions of a flat surface and have to rely on the spatial notions yielded by the eye, based on the projective geometry of retinal images and the laws of optics. And here we meet one of the greatest peculiarities of Western art, i.e., perspective. Great painting traditions have flowered for millennia, innocent of this breakthrough. The discovery of the laws of perspective is one of the glories of the Italian Renaissance; they mirror optical laws easily demonstrated a few centuries later by the darkroom and the photographic camera; they belong then to the naturalistic mood of Renaissance culture. But they also enhance its Platonic spirit, because perspective is, in effect, a glorification of the centre. The whole picture is arranged around a generating centre, toward which all lines converge. Gestalt psychology and the psychology of vision have discovered that the eye naturally organizes images along axes, of which the vertical one, related to gravity, is the most important; centres can emerge only where two or more axes cross-cut (Howard and Templeton, 1966), and it has been shown that the eye automatically orients itself, through saccadic movements, toward such information-rich areas (Kuhn, 1974). Some information theorists suggest that when drawing pictures, the tracts do tend to be organized around an invisible central area (Ceccato, 1968). The idea of perspective is perhaps grounded in psychophysiology, but there is no doubt that its development is a wholly cultural affair. Its uniqueness to Western civilization can be readily seen by comparison between Japanese and European styles of interior architecture and garden landscaping. Chang (1938) has characterized Chinese epistemology as *polyocular,* i.e., interested in cross-perspective relations and complementarities rather than in axial distinctions and oppositions.[6] There is no relevant centre in the well-known graphic symbol of yin and yang.

Eliade (1969: 39) has warned us that "we must not envisage this symbolism of the Centre with the geometrical implications that it has to a Western scientific mind. For each one of these microcosms there may be several centres. . . . All the oriental civilizations had an unlimited number of centres."

## THE UNICENTRISM OF WESTERN COSMOLOGY

Galtung (1971), one of the most energetic promoters of the centre-periphery metaphor in the social sciences, has explored such elements in the opposing "Western" and "Oriental" cosmologies. In a vein similar to C. Kluckohn and F. Kluckohn's analysis of cultural values, he states that cosmologies can be analysed according to their notions of space, time reality, man-man interaction, and man-nature interaction.[7] Starting with the latter one, Galtung sees Western culture as characterized by the Cartesian matter-mind duality and the attribution of individual soul to each man, while Eastern culture is characterized by the conception of mind and soul as a common medium diffused throughout the universe. *Man-man interactions* in the West are individualistic and competitive; in the East they are more consensual and harmonic. *Reality* is approached in the West by the Scholastic drive to clear-cut distinctions and the Cartesian analytical mood: "sorting" (i.e., sharp boundary-making) characterizes the Western treatment of reality. In the East the "holistic" stance predominates, and everything is considered interpenetrated with, complementary to, contradictory with, and inseparable from, everything else.

In the West *time* is conceived as unilinear and progressive; in most Western cosmologies it displays a characteristic "trajectory" marked by a primeval high point or golden age, a subsequent vertical fall from grace, a period of darkness and a long ascending parabola of enlightenment, progress, and final catharsis. On the other hand, the East views time as repetitive, turning into unending circles, spirals, or waves.

Galtung tends to conclude that even the most general cultural traits mentioned above can be boiled down to the ultimate one, the Western obsession with the centre. Western *space* is unicentric, i.e., polarized by one centre from which everything else radiates and is measured, and which posits a negatively defined periphery. On the contrary, Oriental space is policentric. As a leading champion of the cause of the Third World, i.e., world peripheries, and as a sympathizer of countercultural movements, he is of course very fond of what he sees as Eastern cosmology and tends to ascribe to Eastern influences the good things in Western thought, such as dialectics. It is worth noting, at this point, that Galtung's characterization seems closely akin to what a leading scholar of cybernetics—a Japanese-American, not surprisingly—calls "traditional Mainstream Logic" and "Emerging Logic" (Maruyama, 1973).

| **Traditional Mainstream Logic** | **Emerging Logic** |
|---|---|
| unidirectional | mutualistic |
| uniformistic | heterogeneistic |
| competitive | symbiotic |
| hierarchical | interactionist |
| quantitative | qualitative |
| classificational | relational |
| atomistic | contextual[8] |

## POLITICAL CENTRALIZATION

Be that as it may—and such sweeping generalizations are always open to challenge—there seems to be a certain consensus on the importance of the centre of perspective in Western art and culture. Mumford (1961) equates centralistic perspective in architecture and urban planning with the centralization of power in what he calls the Baroque city. Counterreformation Rome was certainly self-consciously restructured as a network of centres or nodal points pregnant with sacred and scenographic values and linked by rectilinear avenues. Renaissance planners produced innumerable projects and some actual cities with circular, radiocentric, and concentric ground plans; Palmanova is perhaps the largest and best known. The epitome of the radiocentric pattern, the purest triumph of perspective, can, however, be found in Versailles and the innumerable imitations by other European rulers and aristocrats, with the *patte d'oie* innerving the whole landscape to the horizon and conveying it, so to speak, to the feet of the lord. Rome and Versailles, the centres of absolute spiritual and temporal power, set the example for cities and villas alike, their fame reaching even the Far East. New cities were designed along these patterns, and old ones—the most macroscopic case being Haussmann's Paris—were renewed by clearing great focal centres and creating perspectives toward them. Centralized power and central perspectives constitute two faces of a cultural syndrome within which it is not easy to impute priorities. Certainly the reconstruction of the land- or townscape in radiocentric patterns is a projection of the central power, a materialization of a sociopolitical structure. But there are also grounds to believe that centralization of power and the reconstruction of social reality into a monolithic pyramid culminating in the absolute authority of the sovereign is a reflection of *esprit de geometrie,* a Cartesian passion for the rationalization of social relationships according to deductive logic, standardization, and simplification. Was it their lust for power that led the French Jacobins to bring to ultimate perfection the centralizing processes initiated centuries before by the monarchy, or was it their passion for "reason," i.e., deductive logic? Again, as in the case of "spontaneous" centralization processes in the growth of settlements, there seems to be no conclusive evidence, and the temptation is great to find the easiest way out by positing a mutually reinforcing causal loop among these phenomena.

The effectiveness of such a syndrome has been indeed enormous, as witnessed by France's human geography and by the strand of correspondingly forceful polemics against centralism that can be found in French political thought from the Fronde liberalists, to Catholic writers of the Restoration, to the libertarian and anarchic socialism of Proudhon, down to the passionate utterings of Simone Weil's political testament, Alexandre Marc's relentless battle for federalism and, more recently, De Rougemont's (1977) plea for regionalism. Such a profusion of intellectual efforts does not seem to have noticeably affected deep-seated French outlooks. Even scholars of peripheral regions seem to share it: two recent such studies open with almost identical words: *''En sortant de Paris par la nationale 4''* (Bonnet, 1972) and *''Quittant Paris par la route par un clair apres midi''* (Kessler and Steinbach, n.d.). There is hardly any doubt on the centre of the French view of the world.

Throughout the centuries, France has set the pattern after which the modern nation-state was fashioned. All over Europe absolute monarchies gave way to democratic republics and, in some cases, were transmogrified in totalitarian tyranny with an uninterrupted crescendo of centralization. Even English thought was affected, mainly through the vast works of Jeremy Bentham, whose drive for reform had a vicious antilocalistic, centralistic orientation, and whose *pan-opticon* principle, quite fittingly, became the basic guideline in the design of jailhouses for at least a century.

According to de Tocqueville's classic analysis, centralization is the necessary consequence of such basic processes as democratization and social equalitarianism. Modern enlightened conservatives like Nisbet (1976) seem to keep their analysis within this framework. But it seems likely that centralistic trends and ideologies are rooted in deeper layers of Western culture, as we have seen. Perhaps more important, political centralization is also an aspect of structural processes pertaining to the operation of large-scale technological systems of production, transportation, and communication. It is of little use to preach for political decentralization as long as a way is not found to offset the advantages of large-scale industries, or to operate huge networks of canals, railways, and highways on a local basis, or to strip of its political relevance the design, construction, and operation of such systems (in the Marxian vein of "administration of things" replacing the "government over men"), or to make do without them, according to the "paraprimitive solution."

In sum, centralization of power is not simply the outcome of a conspiracy of power-hungry demagogues, as Nisbet implies; nor only of the fundamental cultural trait of Western civilization, as Galtung suggests; it is also, and perhaps in a more important way, the result of the working of the basic mechanisms of societal cybernetics.

At this point, however, our discourse must radically shift back from a culturological perspective to a systemic one; from a consideration of centres as static spatial concepts to their role as loci of activities; to be more precise, of *control processes* within *systems*. This will prepare the ground for a discussion of peripheries, the conspicuous missing element in our discussion so far.

## THE FUNCTIONS OF CENTRES IN SYSTEMS

There is some agreement that the centre-periphery metaphor, as used in the social sciences, entails two assumptions: (1) that the centre is the locus of decision-making, i.e., of power; (2) that they both belong to an encompassing system, of which they are differentiated but interdependent parts. This is most clearly seen, for example, in the politicological and organizational use of the concept (Gremion, 1976; Tarrow, 1977). An adequate discussion of these assumptions would then call for analysis of such disproportionately large subjects as (1) social power, (2) social systems, and (3) the relationships between the functional (analytical, normative, symbolic, and behavioral) dimensions of society and its spatial (concrete, matter-energy, and communicational) ones. A clarification of such issues would be in order because of the inevitable tendency to water down the spatial denotations of the centre-periphery concept, and to use it as a mere synonym of other polarities, such as bourgeois-proletarian, dominant-subordinated, developed-underdeveloped, rich-poor, urban-rural, and so on. The inclination is inevitable because of the quite real interplay between the spatial and the analytical (behavioral, and social) aspects of society, for the simple fact that society is *both* a collection of physical human organisms on territory *and* a "reality sui generis," as Durkheim stressed, located as a structure of images, norms, and values in their immaterial minds. Thus a phenomenon such as power has both material referents (e.g., transmitted commands, means of coercion, and resources) and symbolic, mental referents (e.g., consent, fear, legitimization, resistance, and persuasion). Power can be exercised by operating a complex machinery of means of coercion, corruption, and propaganda, but also by letting people embrace the values, understand the words, and anticipate the will of a charismatic leader. Power might be embodied in a specialized communication subsystem, such as the central nervous system of organisms, or may be diffused throughout the system, such as the humoral systems of enzymes.

The sociological literature on power (and related concepts such as authority, control, influence, domination, and so on) is enormous and controversial, since the issue is objectively very complex. Thus we can hardly be expected to achieve here a clear synthesis. The present view comes mainly from Deutsch (1963), Etzioni (1968), and Kuhn (1974). Suffice it to emphasize that the different forms and bases of power have quite different relationships with the spatial configurations of society. On the one hand, coercive power, based on continuous surveillance by the dominant of the subordinate's behavior and on quick deployment of threats and punishments, is clearly tied to material conditions such as efficient communication lines and distribution of coercive "resources" (or "bads" in Boulding's language) on the dominated territory. To the extent that political systems (states and governments) are, by definition, based on the monopoly of the use of force over a territory, they must take into account these hard realities of a physical and spatial nature. Thus state systems tend to look, from a bird's (or geographer's) eye, like unicellular organisms with nuclei, (core, centre, and capital), bodies (the territorial expanse), and outer membranes (the boundary) with stomi (border passes, ports, and so on). In a diachronic perspective, states have often also displayed ameboid changes in body size, shape, and location across space. The organismic metaphor in political geography has the virtues and pitfalls of all metaphors employed by scientists in every discipline and there is no reason to be peculiarly touchy about it. Deutsch (1963) and D. Faston have shown the great potential of what is essentially an updated organismic metaphor—the cybernetic one—in the analysis of political systems. Within this framework, power is defined as control over communication flows, and such control is exercised at the nodes of the channel systems; in particular, at those nodes where *selection* is possible, i.e., where *decisions* can be made. Miller (1965a, 1965b) formally defines decision-making subsystems as nodes with fewer output than input channels.

Channel networks are physical things located in space; thus there are physical loci of power and decision-making. As complex systems usually achieve their complexity through hierarchicisation (Simon, 1969), there emerges a hierarchy of power centres branching out to a seat of ultimate decision-making power. Commands flow down from the centre to the periphery, while information travels in the reverse direction.

In opposition to this communicational-cybernetic, ultimately physicalist concept of power systems, we have the functionalist-normative view of society so majestically represented by the Durkheim-Parsons "grand tradition," where power itself almost disappears, replaced by the concept of collective moral consensus on shared norms and values. Social order is achieved not through the operation of an ultimately coercive (political) power system, but through the spontaneous coordination of individual behavior by means of impersonal mechanisms such as the market. People exercise functions and not power; status and prestige are peacefully and unanimously granted to those who fill the most strategic social roles, for example. This type of society, based on commonality of immaterial elements such as images and moral codes, is obviously much less tied to the physical laws governing channel networks and is also less bound to spatial determinants of communication systems. To speak of centre and periphery in such a society can be wholly misleading. As it has been pointed out by L. Kristof, such societies have "cores" rather than centres; as they grow, any initial difference between core and hinterland is submerged.

Such societies are held together not by a centre of physical control, but by a symbolically central structure of values and norms. Shils (1975: 3) has articulated the clearest statement of this approach to centre and periphery:

> Society has a center. There is a central zone in the structure of society. The central zone is not, as such, a spatially located phenomenon. It almost always has a more or less definite location within the bounded territory in which the society lives. Its centrality, however, has nothing to do with geometry and little with geography. The center, or the central zone, is a phenomenon of the realm of values and beliefs . . . . The central zone partakes of the nature of the sacred . . . . The center is also a phenomenon of the realm of action. It is a structure of activities, of roles and persons, within the network of institutions.

Of course, being a reasonable man, Shils is far from blind to the institutional, organizational, political, and territorial aspects of societies. Thus the centrality of a value system is also due to its being promoted by the ruling elite:

> The decisions made by elites contain as major elements certain general standards of judgement . . . the values [of] which are inherent in these standards . . . we shall call the central value system of the society. This central value system is the central zone of the society. It is central because of its intimate connection with what the society holds to be sacred; it is central because it is espoused by the ruling authorities of the society [Shils, 1975: 4].

In another essay Shils (1975: 80) notices the "territorial delimitation, which, in turn, gives rise to the centre-periphery polarity"; he notes that "ecologically, as the distance from the centre increases, there is a diminution of the effectiveness of integration of all types" and that societies differ in their amounts of centralization and centre-periphery differentiation. But the focus of his approach is unmistakably culturological: From the observation that societies tend to develop spatial centres, he is lead to discover the affinity, already stressed by Eliade and others, of the central with the sacred; from the observation of the falling integration in the peripheries, he is led to study the processes of integration, the topology of consensus.

Real societies are mixtures of both models; the politicological-cybernetic one strongly tied to a spatial pattern of centre-periphery, and the sociological-culturological one, which is much more volatile in its spatial referents. No wonder then that both the "materialistic" model based on power and communication and the symbolic model based on consent and shared meaning have been subjected to analysis through spatial categories such as centre-periphery.

But the perils of adding confusion in a field such as the social sciences, already cluttered with the debris of perhaps hundreds of conceptual models, "dominant" metaphors, "normal" paradigms analytical categories, and so on, is to be considered. I think at this point it would be useful to adopt the elegant formulations of Kuhn (1974), who radically distinguishes between controlled (formal) and noncontrolled (informal) systems. The organism and the polity are examples of the former; the ecosystem and the market of the latter. The former have subsystems that make crucial decisions purposefully affecting the whole system, i.e., systemwide decision makers, control subsystems, or, in von Bertalanffy's (1968) homely words, leading parts. In the second category of systems, such components are lacking and the behavior of the whole system is simply the unintended result of the behavior of its subsystems. The politicological-cybernetic societal model belongs to the first type of system, while the sociological-symbolic belongs to the second. I would venture that the first necessarily has one, and only one, paramount centre, a point in place where ultimately binding decisions are made;it may be a moving point, such as Charlemagne's itinerant court trailing from castle to castle or Air Force One carrying the U.S. President. But as government becomes an ever more complex affair and decision-making requires ever larger amounts of information retrieved from the archives, the point of ultimate decision-making tended, before the advent of the "compunication (i.e., computer + communication) revolution," to cling to a fixed place where the "buck stops": the court, the parliament, the executive buildings, or the capital.

The second type of system—the market, the ecosystem, and society—culturologically seen as the outcome of spontaneous coordination of individual behavior through common mental programmes or "invisible hands," does not necessarily display such punctiform centres. Even the stock exchange has an areal (field), rather than a nodal, spatial structure.[9] They may have a plurality of coordinated centres, a diffuse core area where integration is more advanced than in the marginal area, or no internal differentiation at all.

Further, I would advance that the first type of system tends to strictly control and close its boundaries (Mayhew, 1971; Kaufman, 1974) and thus create peripheries, while the second tends to have more open boundaries and thus less differentiation between core and periphery. This links up with the dialectics of centres and boundaries already discussed in reference to Eliade's writings. There we recalled some insights on defended and nondefended boundaries, and the idea that centre is somehow the consequence of the maintenance and defence of boundaries. This view does not necessarily clash with von Bertalanffy's (1968: 71) principle that progressive centralization is the consequence of progressive segregation (i.e., internal differentiation) of systems, since their empirical referents are different; and both ideas emphasize that "progressive centralization means progressive individuation," i.e., differentiation of the individual system from its environment or, in other words, stabilization and hardening of its boundaries.

## CENTRE-PERIPHERY DUALISM IN ECONOMIC SYSTEMS

The relevance of the centre-periphery polarity has been well demonstrated by Wallerstein's (1974) work on the emergence of the modern world system, in which he contrasts the relative advantages of political empires, i.e., centralized controlled systems bent on closing and defending their boundaries, and the capitalist economic system, unburdened by hierarchical bureaucracies and boundary-maintaining apparatus. His conceptualization of core areas, semiperipheral areas, peripheral areas, and external arena, as well as his developing theory on their dynamics in the making of the modern world system, look indeed like one of the most sophisticated and ambitious examples of the fruitfulness of a centre-periphery approach. In other hands, however, this polar couple may become a mere new, popular catchword by which old-fashioned Marxian classical economics is modernized with some robes of space economics.

As we have seen, the centre-periphery antinomy seems more illuminating in reference to political systems, where centralization processes have proceeded for centuries at ever increasing rates. But it has been first popularized by political economists, or space economists, under the name of polarized growth or "economic dualism." I have already presented elsewhere a "developmental" (i.e., ecosociological) "perspective to centre-periphery relations" and shall not repeat myself. I am fully aware of the structural processes pertaining to the production and transportation subsystems, to the economies of scale, to accumulation, to external economies, and to general principles of location of economic activities, by which the rich tend to grow richer and the poor lag behind, by which larger settlements grow at the expense of smaller ones and show "apoplexy at the centre and anomy at the peripheries." I am also aware of the discussions among economists on the contending theories of "dependencia," "imperialism," "lag," or "vicious circle of poverty." The point is that a centre-periphery approach seems less legitimate in dealing with economic systems than with any others. Punctiform centres are a reality in the political system; in many of them, and especially until the advent of wireless communication and aerial warfare, control centers tended to have central locations with respect to communication networks. In the social normative system, centres may acquire a symbolical identification with the innermost *sancta sanctorum*. But in economies one can speak only of *core areas,* whose location in space usually has nothing to do with geographical or geometrical centres. The metaphor is simply misleading, for more developed areas, on the contrary, are usually found off centre, at the interfaces with other economic systems, along the coasts, or where chance laid natural resources, and so on. The identification of central with developed, and peripheral with underdeveloped, when institutionalized, for instance, in the EEC, has led to some curious situations, such as fairly well-to-do regions located along the margins being officially classified as central and clamoring recognition as peripheral; or, on the contrary, less fortunate pockets of poverty in central areas being called peripheral. At the global level, the centre-periphery metaphor hides the fact that spheres have no "natural" centres, as circles and other flat figures do; what is seen as centre is completely subjective and "historical," as some of the Chapters in this volume emphasize. If centre is equated with "industrialized," "modern," "capitalist," "advanced," and so on, it is quite meaningless to apply it to such a wide and scattered collection of areas.

Two explanations can be advanced for the usage: the first, recalled by L. Kristof, is its statistical origin; the second is its translated usage from the sociopolitical realm. In this latter case, we can as well forego the translation and resume the analysis of the fields in which the notion originally and more interestingly applies. To repeat, processes of "polarized growth" and "dualism" are only too real; both political centralization and sociocultural dominance of symbolic centres have much to do with them, and the study of their interrelations is highly relevant and interesting, as I have tried to discuss elsewhere. However, in itself, the economic realm does not appear a very fit subject for an analysis in terms of centre-periphery. To do so runs the risk of misplacing in economic processes principles that are proper to political and social processes (which are, of course, spatial); in more frank words, to attribute to "capitalist economy" the sins of the centralized nation-state.

The whole discussion of "imperialism" is vitiated by a lack of analytical clarity about what should be inputed to political and what to economic factors (processes, motives, and systems); sometimes the merchants are seen as well-meaning tools in the hands of cunning, aggressive political-military leaders, while other times colonial troops and missionaries are presented as naive instruments of a capitalist conspiracy to penetrate foreign markets. Similarly, economic isolationism and autarky is in turn interpreted as the result of economic interests or political motivations; this can also be said for the relationship between industrialization and militarism. Is the arms race a function of industry's need for expansion, or is industry a servant of political and military goals? Again, the easy way out is to postulate interdependence and mutual causation. But there are many, such as J. Schumpeter in reference to imperialism and De Rougement (1977) in reference to armaments races, that put the greater blame on political factors; in the latter case, the nation-state's ruthless drive for "unity," i.e., "internal homogeneity," and sovereignty, i.e., independence from the environment. To repeat, this means hardening of boundaries, growing individuation, and increasing centralization.

## SYSTEMS AND BOUNDARIES

Generally speaking, people are interested in the centre, the core of things, and neglect the margins. There are, as we have seen, sound logical, physiological, and sociocultural reasons to do so. Philosophers have recommended that we catch the essence of things and forget the marginal aspects; some logicians maintain that too precise definitions of terms and concepts, in order to discriminate marginal cases, are unnecessary and even stifling (Popper, 1969). Sociologists have usually looked at societies as self-contained systems, whose boundary interactions with other societies and the environments are relatively uninteresting (Mayhew, 1971). Modern societies tend to channel the attention of the masses to what happens at their centres. No wonder then that borders, frontiers, and boundaries are hardly mentioned in most textbooks in the social sciences.

The situation is different with some disciplines specializing in the international field; and it is especially different with political geography, which has developed the most massive amount of empirical studies as well as theoretical considerations on this subject. With one limitation, however, the frontiers studied by geographers are almost exclusively those pertaining to the largest political units: nation-states, empires, and their dependencies. Much less studied are the internal boundaries between administrative units and local institutions.

But there is a new mood in boundary studies not to employ the much-abused concept of "revolution." In political geography it concerns the shift from a Holdichian to a Lydian outlook,[10] from a consideration of boundaries as a line of separation between hostile political units to the consideration of borders as regions of encounter and exchange between cooperating neighbor states; from "separatist" to "associative" policies (Kasperson and Minghi, 1969; Gottman 1973; Dorion, 1974; Guichonnet and Raffestin, 1974). In other social sciences the interest in boundaries is a fallout of the systems approach. Boundaries are an integral part of the definition of system (Miller, 1965a, 1965b). Wherever a social group, institution, or organization is conceptualized as a system, it is automatic to look for its boundaries. In fact, the issue of boundaries is one of the *pons asini* of the whole systems approach.

The new approach to boundary studies does not limit itself to geographical boundaries of societal systems. It is interested in the analysis of the spatial boundaries at all systemic levels, and it is also interested in nonspatial boundaries.

Systems are made up not only of material-energetic components, but also of relations and interactions among them and their attributes. In simple mechanical systems these are embodied in material-energetic connections, depending on physical contact and proximity; as the level of complexity grows, the spatial arrangement, which is the primordial informational structure (form = information; Monod, 1970), becomes more fluid and the communication network becomes more extended, involuted, and "ethereal" (Simon, 1969). Thus it is often better to focus on the behavior of the components themselves, leaving the causal network in a "black box," as it were; to focus on the information content and meaning of the messages, instead of the spatial structure of the communication networks. This is often advisable when dealing with sociocultural systems, where the behavior of components, often involving great changes in the material-energetic posture, depends on very thin informational flows—images, ideas, attitudes, perceptions, and expectations (Bertalanffy, 1968)—within and between the individual organisms.

Thus the boundaries of social systems are not only spatial, but also functional; a social system is said to exist as long as its components display certain behaviours, states, and attributes. At the moment its variations exceed certain critical values or norms, the system is said to be stressed, disintegrated or to have become something else (Buckley, 1968). A family, church, corporation, party, and state have a spatial boundary, i.e., a line circumscribing the localities in which their human (and material) components are placed. But they also have normative, functional, or analytical boundaries circumscribing the range of behaviours and attributes stipulated to belong to the system. When we deal with abstract (or action) systems, it is possible to overlook the spatial dimension and boundaries completely; but this is merely a heuristic, methodological device to simplify concrete reality. Some models of the social, political, and economic systems are constructed not with concrete organisms but with roles and "persons," i.e., fictional disembodied characters ("homo sociologicus," "homo politicus," and "homo oeconomicus"). This may be a necessary first approximation model, but, as science develops, the characters must be integrated, and this usually means meeting the physical, spatial constraints of human behaviour (Miller, 1965a, 1965b).

## BOUNDARIES AND FRONTIERS: CONTRASTING SOCIOLOGICAL APPROACHES

The study of analytical, nonspatial boundaries is not foreign to the Parsonian systems analysis, where interesting insights on the interface between the various analytical (sub)systems can be found. But long before Parsons the interaction of spatial and analytical boundaries had been placed at the centre of a full-scale sociological treatise.

*La théorie des frontiéres et des classes* (1908), by the Belgian scholar and social reformer, Guillaume De Greef, is built around the following wide-ranging theoretical assumptions:

> *J'espère avoir démontré dans cette étude que la frontière d'abord homogène et indivise et, du reste, toujours en rapport avec l'état interne de chaque groupe et avec ses relations vis-à-vis des groupes extérieurs, se différencie au cours de l'évolution sociale en une multiplicité croissante de frontières spéciales qui peuvent cependant ètre ramenées à sept espèces de frontières en rapport avec les sept classes de phénomènes sociaux dont j'ai établi le tableau hiérarchique dans le premier volume de mon Introduction à la Sociologie. Cette différenciation croissante des frontières et des groupes respectifs est un facteur important du nivellement des conditions sociales et de l'extension de la civilisation, à condition d'être toujours accompagnée d'une coordination appropriée.*
>
> *La fonction des frontières n'est donc pas simplement séparative; c'est là seulement leur caractère négatif, le plus apparent mais aussi le plus superficiel. La fonction positive des frontières est d'équilibrer les forces de tout groupe social à l'intérieur et d'équilibrerle groupe lui-même vis-à-vis des forces sociales et des groupes extérieurs. Un autre caractère positif et non moins essentiel de la frontière est d'être un organe de la vie de relation intersociale; elle est l'organe de la sensibilité collective aux influences du dehors et en même temps un organe de pénétration du dedans vers le dehors. La vie internationale, la vie mondiale se constituent par l'interpénétration réciproque des frontières, par l'établissement d'un niveau général et commun grâce à cette pénétration et, dans la société ainsi agrandie, par une multiplication croissante des subdivisions intérieures, multiplication qui, dans les sociétés progressives, est accompagnée également d'un nivellement de ces subdivisions.*
>
> *La structure mondiale de l'humanité ne doit donc pas être conçue par nous comme dépourvue de frontières; la planète est limitée, ce serait, dès lors, la seule frontière extérieure en y comprenant la zone susceptible d'être conquise par la science. Mais les frontières intérieures, celles des groupes sociaux particuliers, ne feront que se multiplier sans limite fixe assignable actuellement. Cette différenciation croissante doit nous apparaître comme le procédé naturel de l'évolution progressive de l'humanité* [De Greef, 1908: 386].

De Greef's work was left without any appreciable echo in the history of sociological thought.[11] At the same time, however, wide publicity, at least in the English-speaking world, was given to the other major approach to frontier studies, introduced by F.J. Turner. Sociological studies on *frontiers* in the peculiar American meaning were conducted in Australia, Brazil, South Africa, Indonesia, and other areas (Leyburn, 1970; Katzman, 1975). *Frontier* as the advancing front of Western civilization into "empty" or "wild" areas, for purposes of economic exploitation and cultural civilization, became a basic concept in the language of spatial economy and regional planning (in opposition to periphery, as we shall presently see; Friedmann and Alonso, 1969). Some thought to the role of frontiers as "crossroads" of civilizations was given by the great synthesizers of "sociocultural dynamics" and by world historians. Toynbee (1954) advanced a thesis on the "progressive shifting of power from the centres to the peripheries" as great civilizations degenerate at their core, while some of their elements are assimilated by the "barbarians at the periphery," the "external proletariat" that thus grows strong enough to conquer the centre or break up the old cultural area and build a different one around a new core (the old periphery). Examples of the first process come readily to mind: the Macedonians' conquering of Greece, the Manchu over China, and so on. Examples of the other are also common: the Germanic-Roman sacred empire growing along the old Rhine-Danube frontier of the ancient Roman empire.

An important discussion on these admittedly highly general issues concerns the role of frontiers as crossroads, middlemen, and eventual synthesizer between different sociocultural systems. As an extreme representative of one view we may take the Polish historian Koneczny (1962: 25–26, 318–319), according to whom such processes can only occur between different cultural subdivisions *within* a single civilization:

> There is a principle of closedness, of self-containedness of civilizations. . . . Therefore there is only one law in history: every civilization, so long as it is viable, tries to expand; wherever there meet two civilizations which are able to live, there must be struggle against each other. Every civilization is on the offensive, so long as it is not dying. . . . A synthesis of civilizations does not exist and is not possible. The only thing that is possible . . . is only a mechanical mixture. . . . But its results is only chaos . . . . There are no syntheses, only poisonous mixtures.

The reason on which such sanguine views are grounded is that "such mixtures are a sin against the fundamental condition of every civilization, which is the law of harmony of existential categories. . . . *One cannot be civilized in two different ways.*" Of different opinion are other authoritative world historians; both Toynbee and Sorokin stress the importance of contact areas as crossroads between civilizations, and Braudel (1973: 44) remarks that not only in their core areas do civilizations develop, but also along their contact areas "small sparks can start great fires." The crux of such discussions rests, of course, in the theoretical definition of "civilization," as opposed to "internal subdivisions," and on the empirical identification of such objects; it is on such matters that most world historians and comparative social theorists disagree.

Sociology proper has tended to keep clear of such sweeping generalizations, taking refuge either in purely formal theorizing or in narrow-minded empiricism. "Society," subreptitiously identified with the nation-state, was taken as the largest unit of analysis, and no thought was generally given to higher units as civilizations or international systems. The latter field was usually abandoned to political scientists, while the study of such "ideological" notions as (supranational) "capitalist society" was left to Marxist literature.

Only recently and sporadically have sociologists seriously tackled the problem of interaction between societies, and therefore focused their attention on boundary processes.[12] In one of the clearest and concise books of this new approach (Mayhew, 1971), we find concepts closely reminiscent of De Greef's: Different structural-functional components of the social system have different dynamics processes, limits, and ranges of action. Hence they project very different boundaries around them. The boundaries of the social subsystems are noncoinciding, incongruent, and obey different dynamics. To give two extreme examples, the primary solidaristic groups must be limited in both extension and number of members by the biopsychical limitations of the individual; they tend to be small and to have narrow and closed boundaries. On the other hand, cultural systems, ideas, and ideologies have almost no physical and spatial limitations and can cover the whole globe, leaping over geographical, demographic, and political barriers. Two crucial subsystems occupy an intermediate position: The economic system, which, unlike the primary group, is not bounded to its own penetration and expansion in the environment, even if it does not have all the fluidity of the noosphere; and the political system, whose main function seems to be that of trying to make the various boundaries of the subsystems coincide toward the outside, i.e., to unify and control them in order to be able to put up more resistance to environmental variability and to attain the system's ends.

The upshot of Mayhew's (1971: preface) analysis is this: "It is the overlapping character of the boundaries of our social systems that explain much of the tension and the dynamics of social life." With these statements by an authoritative theorist, the study of boundary is placed right in the centre of sociological concerns; but it must be admitted that no noticeable effects have ensued in day-to-day sociological activities. This discipline still lags behind most others, especially political geography and political science, in the actual study of boundaries, borders, and frontiers, as a review of the literature has shown (Strassoldo, 1977). The subfields most active in such research are human ecology, ethnic studies, and organization studies.

As the most spatially oriented of sociological subfields, human ecology has dealt with boundaries since the Chicago days, when the main problem was to *define*, i.e., to draw and study, the "natural areas" within cities. Also the special field of "community studies" has been concerned with the problem of finding the boundaries around human communities, both urban and rural. One of the most brilliant recent examples of the Chicago tradition is G. Suttles's (1968) study, in which neighbourhood boundaries are shown to have a reality sui generis, independent of other sociocultural factors such as ethnicity and class. This study is influenced both by symbolic interactionism and labeling theory (the boundary as the outcome of interaction between the community and the environment), and by Lorenzian ethology (boundary as the limit of territory).

In ethnic studies the boundary comes up in two ways. On the one side, as the symbolic, normative boundary that keeps ethnic groups apart and defines the relations between them (Barth, 1969; Hughes, 1971). On the other side, territorial boundaries or border zones between ethnic groups and cultural areas are usually a place where contacts and sometimes mixtures occur; often the border areas of nation-states are also marginal and backward places, where ethnic differentials are better preserved or acquire dynamics of their own. For all such reasons, as well as for the case of comparative analyses, border areas are favourite places for sociologists and anthropologists interested in ethnic problems (Rose, 1935; Surace, 1969; Miroglio, 1969; Cole and Wolf, 1974; Gross, 1978).

Finally, organization studies have been led by the influence of general systems theory to focus on the processes and problems of analytical or normative boundaries of administrative or business organizations.

## MARGINALITY

A body of literature not unrelated to the issues mentioned before is that of the "marginal man" (Stonequist 1937), the "member of two worlds," who in some ways belongs to one system and in some ways another. This may be conceptualized as a case of role conflict, cognitive dissonance, and incongruence between membership and reference groups of crosscutting role sets.

Such situations are usually seen as straining and challenging; if the challenge is not met, alienation and disorganization may ensue; if it is met, there is opportunity for creative syntheses. Marginality thus seems the individual, psychological parallel of the frontier situation. The marginal man may work toward a synthesis of the two systems on whose contact area he is situated. The builders of the EEC came from border regions: Adenauer from the Rhineland, Schumann from Alsace, and Degasperi from Trentino. On the contrary, he may strive to improve his position relative to the one centre he feels most attached to. This may in turn lead him to overstress his loyalty in order to overcome his frustration and minority complex. It has been noticed that many great empire builders (and empire-building has often been considered a symptom of internal disorganization) came from areas marginal to great culture cores: Alexander from Macedonia, Tamerlane from Mongol-Turk frontier region, Napoleon from Corsica, Stalin from Georgia, and Hitler from Austria (Devereux, 1973); on a more modest level, we may add Garibaldi from Nice.

More recently *marginality* has increasingly been used in sociological parlance simply as a combined synonym of weakness, poverty, and ignorance to indicate—in a meaning taken over from economics—a position of deprivation (alienation) with respect to power, wealth, and culture (Germani, 1975). No precise, consistent, and relevant spatial denotations seem implied in this usage, although marginal people are usually found in particular ecological pockets within city systems, or in certain rural and peripheral regions within societal systems. The usefulness of the aggregate term and its spatial suggestions has yet to be ascertained, as we shall see in the closely related, almost synonymical, case of "periphery."

## THE PERIPHERY AS A SPECIAL BORDER SITUATION

The relevance of all this for the centre-periphery theme has already been mentioned. An emerging "general theory of boundaries" (Strassoldo, 1973, 1975, 1977) as a subfield of the general theory of systems, likewise fashioned out of a wide variety of disciplinary contributions, results, among other things, in a typology of "border (or boundary) situations."[13] The typology is built upon the contrast between frontier and periphery formalized in the literature of space economics and planning, but also widely used, perhaps less precisely, in other fields and in common parlance. According to Friedmann and Alonso (1969), frontiers are areas of growth in "virgin" territory, while peripheries are the stagnant, quasi-colonial areas beyond the centres or metropolitan regions.

The study of border areas suggests that the closure of boundaries tends to discourage investments, cause higher costs, and generally have depressing effects on economic activities, while open boundaries and energetic interaction with the environment tend to attract people and capitals (Romus, 1971; von Malchus, 1975). The peripheral situation is then associated with closed boundaries, while the frontier situation is related to open boundaries.

This, however, is not enough. Societies are characterized not only by the openness or closure of their boundaries, but also by their spatial *mobility* or *stability*. Older political geographers used to give much attention to the changes in shape and size of political units and to boundary dislocations; the frontier was even sometimes defined as an organ of penetration and attack of states upon each other. All this is now wholly *passè* with the "freeze" on territorial changes tacitly agreed upon by the international community since World War II. But the expanding frontier has been a very important reality throughout history, so the mobility-stability dimension is not illegitimate. The old frontier was open to both the natural and sociocultural environment and to marching; it was a locus of important cultural change and synthesis, as well as of socioeconomic growth. Similar processes can now take place along boundary lines that, although open, cannot be spatially moved. This gives rise to a situation analogous to that of a harbour, a bridge, or a crossroads (Gottmann, 1973), where interactions between different areas take place without changes in their respective locations and without the loss of their diversity and separateness. Indeed, it is just such diversity that creates the "difference of potential" from which the interactions flow. If the boundary could move, the area would be homogenized and the opportunities for exchange lost; lively frontier outposts would become sleepy backwaters.

To round out the typology, we can speculate what the opposite situation—a closed but dynamic boundary, i.e., a moving periphery—might be. This is the "no-man's land," or, in extreme cases, the politics of "scorched earth." In the case of *expansion,* the enlarging system is not interested in interaction with the social or natural environment; the only conceivable reason might be the pursuit of isolation and security, which could entail the creation, around the settled territory, of empty areas only surveyed by mili-

TABLE 2.1 A typology of border situations

| | | *Spatially* | |
|---|---|---|---|
| | | *Dynamic* | *Static* |
| Socially | Open | Frontier | Crossroads |
| | Closed | Scorched earth | Periphery |

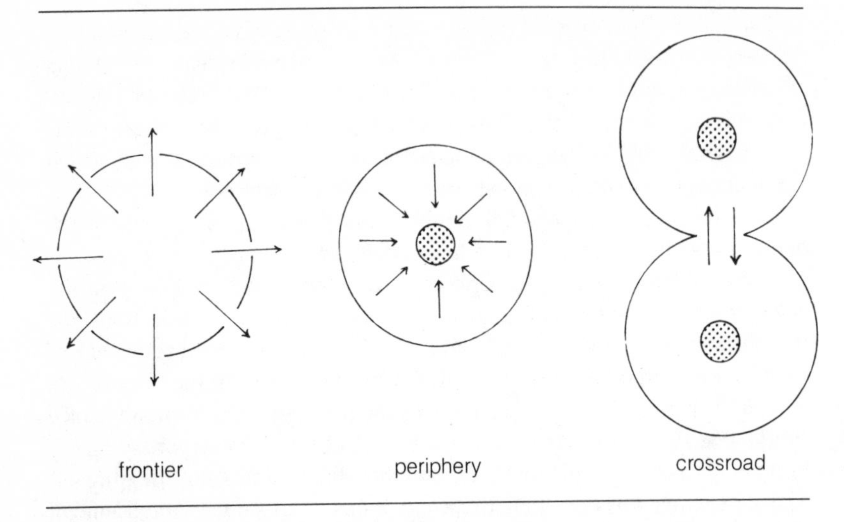


FIGURE 2.1

tary forces. Historically this has often meant the expulsion or destruction of autochthonous populations, the prohibition of cultivation and settlement, and in some extreme cases even the burning out or cutting of vegetation (Miroglio, 1969). This is the real "no-man's land." In the case of *retreating* periphery, we have the classical "scorched earth" policy by which any basis for interaction with the advancing hostile system is destroyed.

The periphery then is an area characterized by having its "shoulders to the wall" of a closed and frozen boundary. It is poor because it is outlying and handicapped by greater costs of transportation to the centre, by which it is forsaken. It is marginal, weak, alienated, provincial, and backward. Innovations come late and investments tend to have a purely exploitative and colonial character, with no spin-off effects because of lack of locational advantages other than natural resources. Its peculiar function in the system is to play host to boundary-maintaining personnel; military bases and garrisons constitute an emerging feature of its cultural, political, economic, and urbanistic makeup. But military presence, while activating certain economic circuits in the tertiary sector, acts as a deterrent to investment in civilian public works, infrastructure, and production (Strassoldo with Gubert, 1973).

Such is the standard picture that emerges from the literature of "border regions" flowering within the EEC countries;[14] but the picture does not seem very dissimilar to that of former colonial territories—the overseas peripheries of empires. Peripheral regions within nations begin to be called "internal colonies" (Hechter, 1975). Thus it is no wonder that the centre-periphery concept has been stretched at the global scale. What is really interesting from a culturological point of view, and that could hardly have been brought to light without subjecting the matter to an analysis in terms of "general theory of boundaries," is that the periphery concept entails the concepts of boundary closure and fixation. This is the specular opposite of the concept of frontier, which has so long dominated Western social science and the Western world view.

The earth was then conceived as an open system, inexhaustible of opportunities to be seized, boundless in resources to be exploited, and limitless in its potential for growth. Mankind seemed launched into a unilinear, infinite evolutionary trajectory; civilization would expand all over the world and pull laggard peoples to the light of progress. Advanced countries felt the responsibility of "the white man's burden" and poverty, ignorance, and barbarism seemed a passing condition. In sum, the whole earth beyond Western countries was an open frontier with no limits to development and no boundaries to the future.

How different is our present mood. We are besieged by the Cassandras of the limits to growth. We count in decades the duration of our nonrenewable stocks of basic resources. The space adventure has shown, if anything, the impracticability of colonization of outer space, at least within meaningful deadlines. No one believes any longer in unilinear theories of evolution and few in the necessity of progress. The world is full and wholly cut up. There are hardly any surprises to be expected in the still underexploited regions. There are no open frontiers, only closely guarded boundaries everywhere.[15] Worse, development does not seem to be automatic. On the contrary, vicious circles of poverty seem to have been ignited by which the expectations of poorer regions to follow the path of the more fortunate ones seem utterly frustrated. Worse still, according to a common view, the continuation of development of the advanced countries necessarily requires the plight of the poor ones. While the peripheries of the centres, i.e., the exploited classes and internal colonies of the advanced countries, could expect an improvement of their position as a consequence of overall economic growth and/or territorial expansion, such perspectives are utterly foreclosed to the world peripheries. There is no "outside" to expand into, and into which to project hate as well as hope. The world is a closed system and its "contradictions" are "exploding" inside it. Strains and conflicts cannot be exported. There is no frontier as outlet for the frustrations produced by social inequalities. The world centre—the most fortunate citizens in the community of nations—cannot quiet the demands of the peripheries with vague promises of future gains, nor can they be advised to "go west, young man." Thus the issue of redistribution of presently available scarce resources cannot be eluded with all that this means in terms of power, prestige, and class conflicts. This is, I think, the doctrine hidden in the centre-periphery concept and the cultural mood that explains its popularity.[16]

## CONCLUSION

The weakness of culturological analyses is their lack of predictive power. Their subject matter are meanings, moods, attitudes, images, and ideas, all of a very volatile nature.

It might well be that in a very short time the centre-periphery concept will be discarded as wholly irrelevant to changed conditions or changed perceptions of them. The history of social thought is a huge attic of obsolete analytical instruments and ill-conceived inventions.

The concept of periphery, as we have seen, is quite young in the social sciences; it is much younger than the opposite one of frontier, which nowadays seems so sadly outdated or objectionable, with its overtones of colonial exploitation. Certainly the idea of centre is infinitely older and more deeply rooted in the human mind; but, as human beings learn to make do without the idea of God and the sacred (and perhaps war and enemy) that seemed so indispensable to their mental operations, they might also outgrow such fixations and come to regard them as childish; they might learn to live in a world without a centre or with a large number of them arranged along sectoral, geographic, or hierarchical lines. They may learn to master their biocultural heritage and their physiological constraints, and get accustomed to looking at the world as an unfocused environment and at society as a community of equals, without kings, presidents, or capitals to symbolize their unity.

The centre—and, to a lesser extent, the periphery—may be a category of the mind of almost Kantian status, at least in Western civilization; but as we are disposing of other fundamental categories such as space, time, and cause, we may do the same with centre. Like those others, it was a mental device to simplify the chaos of impressions, a point of departure and reference of perceptions. But simplicity, like economy, parsimony, and the like, is a requirement of *human* mental operations. "Artificial intelligence" can help us handle and control complexity; a cybernetic society shall be less bound to the mythology of the centre; its integration can perhaps be achieved through diffuse networks of interconnected communication systems.

The "global village" can resemble the villages of old if it will be able to recreate such collective, segmentarian, nonhierarchical structures.

There was no meaningful centre-periphery differentiation in primitive, "mechanical" societies of small, self-sufficient units. That polarization began with the emergence of urban civilization and the network of central places and reached its apex in industrial, capitalist societies where economies of scale, external economies, economies of agglomeration, and so on are responsible for large concentrations of men, capitals, and settlements. Most of these sorts of economies depend upon the technology of transportation and communication, although they can be reinforced by social values and political-military considerations. But technologies change, and so do social values and these other factors.

The cost of communication and transportation has been dropping very steeply, so that it is common nowadays among urbanists and planners to speak about "frictionless space" and locational indifference. To be sure, space and distance will always play a role in the structuration of human activities, but less and less the determining role. So there will be increasing leeway for industrial and other activities to settle in peripheral locations. There may even be more convenience to do so, in terms of available space, freedom from congestion, and so on. The question is whether the higher economic functions of management, finance, research and development, and the like can alsodeconcentrate; but few rational economic factors militate against decentralization. The technology of communication works in this direction. Centralization of headquarters is often a matter of mere prestige and tradition, but other environmental factors, such as residential amenity, seem to be working fast in the opposite direction.

So it seems that the postindustrial, communicational, informational, "technetronic," cybernetic society toward which our civilization is moving, according to many of its most perceptive observers, has no major objections to the end of the centre-periphery duality.

Politically and militarily, there seems to be positively no value in the centre-periphery arrangement. On the contrary, concentration is a military liability. Under the pressure of the atomic missile threat, the gospel is decentralization, mobility, small scale, self-sufficiency. On the political side, the situation is more complex. The political power of big urban concentrations and developed regions is usually not commensurate to their economic strength: Wealth does not translate directly into power. As we have seen, the political systems strive for concentration of *control,* not necessarily of things and people. If better integration can be achieved by territorial equilibrium, the development of the peripheries and local autonomy, it will try to grant it.

So we come to social and cultural considerations. On one side we see inequality, whether economic, social, legal, or territorial, as universally rejected with increasing indignation. The spread of this value translates operationally at the political-economic level in an increasing intervention of the central state through regional planning processes to develop the depressed, peripheral regions. On the other side, we see the charm of the centre being dispelled every day. The centre was a sacred notion, while society is increasingly secular; its glorification is connected with the religious-artistic-cultural apparatus of central power holders. The lavish construction of ceremonial centres and national capitals was an aspect of the attempts of central powers to fascinate and integrate the peripheries (as much as the foreigners), to lure attention and prestige, to impress with artistic achievements, and to buy political loyalty through the manipulations of minds.

All this is nowadays fairly debunked intellectually, even if still going on in practice; the emperor's clothes have been stripped to show his naked power. Monuments are attacked as a waste of resources, and most artists and intellectuals refuse to serve, at least consciously, on the staff of the central power. "National" cultures are decried as mass cultures, the abuses they have grown out of exposed, and local, provincial, and marginal minority cultural traditions revived. The quest for local automony and regional equalitarian planning fuels the rebirth of local cultural traditions long repressed into the vernacular limbo.

Another sociocultural trend decries large-scale, bureaucratic, mass society, with its hierarchical arrangements and its centralization, and looks to its breakdown into small, simple, participatory, natural communities; and this certainly is something that is not going to reinforce centre-periphery differentiation.

To sum up, the question is whether the equalitarian trends are powerful enough to offset the polarizing factors. On both sides we find a mixture of sociocultural-ideological forces and technical-economic ones. It is important, however, to note that political factors seem relatively neutral on the issue, that military factors are decidedly against polarization, and that within the economic system there are contradictory forces. What weighs in favor of accumulation at the centre and abandonment of the periphery are mainly traditional patterns, which are sunk investments. Capital invested in a centralized transport, communication, and settlement systems can be a strong motive for further investment in it. Such conservative attitudes, however, are often rational only in the short run and from the individual viewpoint. From a long-term, collective perspective, a radical decentralization might yield higher payoffs.

Finally we can emphasize that the political system's willingness to allow for the development of the peripheries, local autonomy, and so on can only be expected as long as this does not impair the centre's control over the crucial integrative institutions, such as the nodes of ideological manipulation, the higher law-making bodies, and the like. Beyond that there is only the political system's suicide, which is probably a good way out, but hardly conceivable.

All this seems applicable to advanced societies. The situation in underdeveloped countries is different, because (1) the cleavage between centre and periphery, or better, between primate city and hinterland, is much wider, (2) political systems are often engaged in the initial steps of nation self-building, and (3) the resources are so much below the goals that strict economic efficiency must be adhered to. It is therefore possible that for some time the basic problem of such countries will be the development of the centre, while the periphery will be left to wait. However, it is also possible that the Western model of economic development will be abandoned in favor of another one that tries to bypass the urban-industrial accumulation-concentration phase and heads directly toward a balanced growth of the rural communitarian hinterland, while societal integration is achieved mainly at the symbolic-ideological level. The success of this model cannot be predicted, since there are as yet no clear precedents.

## NOTES

1. Some paragraphs of this chapter have been borrowed from Strassaldo (1977, forthcoming).

2. The citations are from an Italian secondary source, as it was not possible to locate the original.

3. The division between "we" and "they," the internal and the external, seems a universal cultural trait; see Levi-Strauss (1966), Benedict (1934), Mead (1964), and Sopher (1972). The depth-psychological bases underlying it have been explored by Freudians such as Fromm (1955) and Fornari (1969). Max Weber commented on the division of internal and external as a basic category of the political realm, connected with its territorial dimension, which in turn is due to the physical nature of violence; but it was K. Schmitt who in 1932 linked such insights with the Freudian concepts. That the primary characteristic of political space is closure or boundedness, in opposition to "open spaces" or "fields," has also been remarked by theoretical geographers: see Cox (1972) and Claval (1973). To sum up with a quotation from a world-historian: "The point of any ethnocentric world image is to divide the world into moieties, ourselves and the others, ourselves forming the most important of the two. To be fully satisfying, such an image of the world must be at once historical and geographical" (Hodgson, 1968).

4. To Vickers (1970: 136), the importance of national frontiers in shaping human life is far from declining, even if their emotional overtones may be changing: "Next to the human skin, [the nation state] is today the most important interface between internal and external relations and I think it is bound to remain so and even to increase in importance."

5. That a basic distinction between "defended" and "nondefended" system-boundaries can be drawn, with far-reaching implications, is suggested by Boulding (1970). De Greef (1908)

emphasized the priority of the boundary over the centre in the formation of systems. Ashby (1962) seems to suggest the same view: a system first emerges as a differentiation between an inside and an outside; the control centre then develops as a response to boundary-maintenance requirements. An elaborated theory of the growth of urban-political systems based on the same principles has been developed by Di Sopra (1975).

6. Citation borrowed from Maruyama (1973).

7. Galtung's views were presented during a seminar at the Dubrovnik Inter-University Centre in November 1976. I apologize for possible misinterpretation or premature schematization of the results of his research, probably still in progress.

8. The reader is referred to Maruyama's article for a fascinating clarification of the antinomies.

9. The formation of prices in a market system, of which the stock exchange is a telescoped example, requires bargaining among a *plurality* of buyers and sellers and hence an area to accommodate them and a space to support a web, however immaterial, of interactions. On the contrary, political decisions—of which administered prices can be an example—require in principle no more space than exists between the ears of the decision maker. But one may well contend that such space, punctiform from the point of view of the social scientist, is a hugely extended, differentiated, complex, diffuse, redundant network from the point of view of the neurophysiologist. It can even be better conceived as a field than as a hierarchical network. No paramount centre has yet been localized in the brain system.

10. Peattie (1970: 55) contrasts the "realistic and strategic" school of thought on boundaries represented in the early twentieth century by British Colonel Holdich and the "abstract and theoretical" school of Professor Lyde. "The first considers the problems of the moment, the second seeks solutions for the future and is more social."

11. To my knowledge the only acknowledgement of De Greef's theory of frontiers can be found in Barnes (1942).

12. A few pages on the different boundaries of societies appear in a treatise by Janne (1968), another Belgian sociologist, but there is no hint of the author's awareness of De Greef's work.

13. As declared in the introduction, neither terminological questions nor formal definitions are considered in the present chapter, as they have been discussed elsewhere. The difference between "border," which has an areal connotation, and "boundary," which is a line, however relevant in other contexts, is of no weight here. Both terms are taken to refer to the more general concept, of which "frontier" and "periphery" are polar subcategories. Different antinomies have been explored by other authors, such as the one between frontier (area of expansion of dynamic societies) and boundaries (linear limits of political-territorial systems; Kristof, 1959).

14. Besides Romus (1971) and von Malchus (1975), see the documents of the Council of Europe from the several meetings on the problems of frontier regions.

15. The closing of the world's frontier has been announced by Mumford (1944). Its social, political, and economic implications have been variously emphasized by Herz (1959), Hoffman (1960), Haas (1968), Mayhew (1971), and Taylor (1973).

16. The concern for world peripheries, so widespread especially among counter-cultural movements, has certainly been boosted by the modern phenomena mentioned above, as well as by the power of mass media, the facility of travel, and so on; but it also incorporates a very long tradition of eulogies of faraway "primitives" as more natural, innocent, and virtuous than one's own immediate fellows. This tradition can be seen in the nostalgia for rural life chanted by urban poets throughout the millennia, from Hesiodus to Virgil to nineteenth-century romantics; in the praises of the courage and hardiness of "barbarians," uttered by people who, like Tacitus, felt that their own society was corrupt; and in the Rousseauian celebration of the "noble savage," as well as in contemporary populistic doctrines. One of the more elaborated social

theories based on the systematic opposition of the civilized urban centres, prone to degeneration, and the barbarian but virtuous rural peripheries, whence reinvigorating waves of regeneration come, was proposed a few centuries ago by Ibn Kaldhoun.

# REFERENCES


ARDREY, R. (1966) The Territorial Imperative. New York: Atheneum.

ASHBY, R. (1962) "Principles of self-organizing systems," in V. Forster et al. (eds.) Principles of Self-Organization. New York: Pergamon.

BARNES, K. E. [ed.] (1942) An Introduction to the History of Sociology. Chicago: University of Chicago Press.

BARTH, F. (1969) Ethnic Groups and Boundaries—The Social Organization of Cultural Difference. Oslo: Universitetsforlaget.

BENEDICT, R. (1934) Patterns of Culture. Boston: Houghton Mifflin.

von BERTALANFFY, L. (1968) General System Theory. New York: George Braziller.

BONNET, S. (1972) Sociologie politique et religieuse de la Lorraine. Paris: Colin.

BOULDING, K. E. (1970) Economics as a Science. McGraw-Hill.

BRAUDEL, F. (1973) Scritti sulla storia. Milan: Mondadori.

BUCKLEY, W. [ed.] (1968) Modern Systems Research for the Behavioral Scientist. Chicago: Aldine.

CECCATO, S. (1968) Cibernetica per tutti. Milan: Feltrinelli.

CHANG, T. S. (1938) A Chinese Philosopher's Theory of Knowledge. Yenching Journal of Social Studies 1.

CLAVAL, P. Principes de geographie sociale. Paris: Genin.

COLE, J. W. and E. R. WOLF (1974) The Hidden Frontier—Ecology and Ethnicity in an Alpine Valley. New York: Academic.

COX, K. R. Man, Location and Behavior. New York: John Wiley.

De GREEF, G. (1908) La Théorie des frontières et des classes. Brussels: Larcier.

De ROUGEMONT, D. (1977) L'avenir est notre affaire. Paris: Stock.

DEUTSCH, K. W. (1963) The Nerves of Government. New York: Macmillan.

DEVEREUX, G. (1974) "La psicanalisi e la storia. Una applicazione alla storia di Sparta," in F. Braudel (ed.) La storia e le altre scienze sociali. Bari: Laterza.

Di SOPRA, L. (1975) Lo spazio merce. Venice: Marsilio.

DORION, H. [ed.] (1974) Les frontières politiques. Cahiers de Géographie de Québec 43, 18.

ELIADE, M. (1969) Images and Symbols. New York: Search.

ETZIONI, A. (1968) The Active Society. New York: Macmillan.

________(1949) Le mythe de l'éternel retour. Paris: Gallimard.

FORNARI, F. (1969) Dissacrazione della guerra. Milan: Feltrinelli.

FRIEDMANN, J. and W. ALONSO [eds.] (1969) Regional Development and Planning. Cambridge, MA: MIT Press.

FROMM, E. (1955) The Sane Society. New York: Holt, Rinehart & Winston.

GALTUNG, J. (1971) "A structural theory of imperialism." Journal of Peace Research 8, 2.

GERMANI, G. (1975) "Aspetti teorici e radici storiche del concetto di marginalitá con particolare riguardo all 'America Latina," in G. Germani (ed.) Urbanizzazione e modernizzazione. Bologna: Il Mulino.

GOTTMANN, J. (1973) The Significance of Territory. Charlottesville: University of Virginia Press.

GREMION, P. (1976) Le pouvoir périphérique: Bureaucrates et notables dans le systeme politique français. Paris: Seuil.
GROSS, F. (1978) Ethnics in Borderland. New York: City University Press.
GUICHONNET, P. and C. RAFFESTIN (1974) Géographie des frontières. Paris: Presses Universitaires de France.
HAAS, E. B. (1968) Beyond the Nation-State. Palo Alto, CA: Stanford University Press.
HECHTER, M. (1975) Internal Colonialism: The Celtic Fringe in British National Development, 1536–1966. London: Routledge & Kegan Paul.
HERZ, H. (1959) International Politics in the Atomic Age. New York: Columbia University Press.
HODGSON, M. G. S. (1968) "The interrelations of societies in history," in L. Kriesberg (ed.) Social Processes in International Relations, a Reader. New York: John Wiley.
HOFFMANN, S. (1960) Contemporary Theory in International Relations. Englewood Cliffs, NJ: Prentice-Hall.
HOWARD, I. P. and W. B. TEMPLETON (1966) Human Spatial Orientation. New York: John Wiley.
HUGHES, E. C. (1971). The Sociological Eye: Selected Papers. Chicago: Aldine-Atherton.
JAFFE, A. (1976) "Symbolism in the visual arts," C. G. Jung (ed.) in Man and His Symbols. New York: Dell.
JANNE, H. (1968) Le systeme social, essai de theorie generale. Brussels: Editions de l'Institut de Sociologie de l'Université Libre.
JOLLY, A. (1972) The Evolution of Primate Behavior. New York: Macmillan.
KASPERSON, E. R. and J. V. MINGHI, [eds.] (1969) The Structure of Political Geography. Chicago: Aldine.
KATZMAN, M. T. (1975) "The Brazilian frontier in comparative perspective." Comparative Studies in Society and History 17, 3: 266–285.
KAUFMAN, G. (1974) Il sistema globale—immagini e modelli. Udine: Del Bianco.
KELLER, R. T. (1976) "Boundary-spanning activities and employee reactions: an empirical study." Human Relations 29, 7.
KESSLER, S. and H. STEINBACH (n.d.) Le coin. Paris: Comité de Défense des Travailleurs Frontaliers de Haut-Rhin.
KONECZNY, F. (1962) On the Plurality of Civilizations. London: Polonica.
KRISTOF, L. (1959) "The nature of frontiers and boundaries." Annals of the Association of American Geographers 49.
KUHN, A. (1974) The Logic of Social Systems. San Francisco: Jossey-Bass.
LEYBURN, J. G. (1970) Frontier Folkways. Hamden: Archon.
LEVI-STRAUSS, C. (1966) The Savage Mind. Chicago: University of Chicago Press.
von MALCHUS, V. (1975) Partherschaft an Europaeischen Grenzen—Integration durch grenzuberschreitende Zusammenarbeit. Bonn: Europa Union Verlag.
MARUYAMA, M. (1973) "Symbiotization of cultural heterogeneity on scientific, epistemological and esthetic bases." General Systems 18.
MAYHEW, L. (1971) Society: Institutions and Activity. Glenview, IL: Scott, Foresman.
MEAD, M. (1964) Continuities in Cultural Evolution. New Haven, CT: Yale University Press.
MILLER, J. G. (1965a) "Living systems: basic concepts." Behavioral Science 10, 3.
———(1965b) Living systems: structure and process. Behavioral Science 10, 4.
MIROGLIO, A. (1969) "Réflexions sur l'importance des frontières des états et des ethnies," in Kontakte und Grenzen. Goettingen: Schwartz.
MONOD, J. (1970) Il caso e la necessità. Milan: Mondadori.

MUMFORD, L. (1961) The City in History. New York: Harcourt Brace Jovanovich.
________(1944) The Condition of Man. New York: Harcourt Brace Jovanovich.
NISBET, R. A. (1976) Twilight of Authority. London: Heinemann.
________(1967) The Sociological Tradition. London: Heinemann.
PAILLARD, J. (1974) "Le traitment des informations spatiales," in De l'espace corporel a l'espace ecologique. Paris: P.U.F[e].
PAULET, G. (1961) Les metamorphoses du cercle. Paris: Plon.
PEATTIE, R. (1970) Look to the Frontiers—A Geography for the Peace Table. Port Washington, NY: Kennicat.
PIAGET, J. (1948) La representation de l'espace chez l'enfant. Paris: Presses Universitaires de France.
POPPER, K. (1969) Conjectures and Refutations. London: Routledge & Kegan Paul.
ROMUS, P. [ed.] (1971) Les régions transfrontalieres a l'heure du Marchè Commun. Brussels: Presses Universitaires.
ROSE, W. J. (1935) "The sociology of frontiers." Sociological Review 27, 2.
SHILS, E. (1975) Center and Periphery. Chicago. University of Chicago Press.
SIMON, H. A. (1969) The Sciences of the Artificial. Cambridge, MA: MIT Press.
SOPHER, D. (1972) "Place and location, notes on the spatial patterning of culture." Social Science Quarterly.
STONEQUIST, E. (1937) Marginal man. New York: Russell and Russell.
STRASSOLDO, R. (forthcoming) "A developmental perspective to center-periphery relations," in A. Kuklinski (ed.) Strategies of Polarized Development. The Hague: Mouton.
________(1977) "The study of boundaries: a systems-oriented, multidisciplinary, bibliographical essay. Jerusalem Journal of International Relations 2, 3: 81–107.
________(1975) "The systemic region," in Les régions transfrontalieres de l'Europe. Geneva: Association des Instituts d'Etudes Europeennes.
________"Regional development and national defense: a conflict of values and power in a frontier region," in R. Strassoldo (ed.) Boundaries and Regions: Exploration in the Growth and Peace Potential of the Peripheries. Trieste: Lint.
________(1970) "From barrier to junction—towards a sociological theory of borders." (mimeo)
________with R. GUBERT (1973) "The boundary: an overview of its current theoretical status," in R. Strassoldo (ed.) Boundaries and Regions: Exploration in the Growth and Peace Potential of the Peripheries. Trieste: Lint.
SURACE, S. (1969 "Per una sociologia delle frontiere: il caso Stati Uniti-Messico." Rassegna Italiana de Sociologia 10, 2.
SUTTLES, G. D. (1968) The Social Order of the Slum. Ethnicity and territory in the Inner City. Chicago, University of Chicago Press.
TARROW, S. (1977) Between Center and Periphery: Grassroots Politicians in France and Italy. New Haven: Yale University Press.
TAYLOR, A. (1973) "Some political implications of the Forrester world system model," in E. Laszlo (ed.) The World System—Models, Norms, Variations. New York: George Braziller.
TOYNBEE, A. (1954) A Study of History, Vol. 8. New York: Oxford University Press.
TREVARTHEN, C. B. (1968) "Two mechanisms of vision in primates." Psychologische Forschung 31: 299–337.
VICKERS, G. (1970) Freedom in a Rocking Boat. Harmondsworth: Penguin.
WALLERSTEIN, I. (1974) The Modern World System. New York: Academic.

# 3

## CENTRE/PERIPHERY AND SPACE: MODELS OF POLITICAL GEOGRAPHY

Paul Claval

### THE ROLE OF SPACE IN POLITICAL THINKING

The concept of space is present from the sixteenth to the eighteenth century in the works of such French authors as Jean Bodin and Montesquieu—but totally absent among their English contemporaries, such as Hobbes and Locke. For the author of *L'Esprit des Lois,* diversity of environment takes the variety of political systems into account. From a rationalist point of view, there is indeed nothing that justifies the multiplicity of forms of government, except for seeing in it a reflection of the environment. Montesquieu adopts that position, although it raises problems that he recognizes: He oscillates between a strict determinism—he needs that to take the multiplicity of political forms into account—and a possibilism that satisfies his sense of human freedom and responsibility.

For the rationalist and evolutionary philosophies, space is not a pertinent element of explanation: When one observes the varying forms in different places, one sees that these places belong to successive moments in history. This brings with it the almost total disappearance of the reflections on space from Condorcet to Marx.

When, in the first political studies, space is taken into consideration, it is, as with Montesquieu, considered as a natural environment. Some authors—this is true for Marx—are well aware that it represents an impediment to transparency, but the idea has hardly been exploited.

Space reappears in the political geography at the end of the nineteenth century. It is conceived as being environmental: From a Darwinist point of view, one tries to measure its influence on human behaviour. It is present in that form in Ratzel's work and in that of most authors who believe in the game of natural selection.

Space is also analyzed as the composing factor of strategies of power that develop in the world. Mackinder and Admiral Mahan put the imperative strategies first, the part played by continental areas and the seas, and suggest an interpretation of the balance in the world in terms of Heartland and Rimland, which also evokes the topical theme of the centre and the periphery. Most experts in political geography, a subject in the process of creation, conceive space rather as a support for economic power; that is very clear in Bowman, but is also the case with André Siegfried. For Ratzel and for German geopolitics, which gets its inspiration from him in this respect, space also intervenes as an amalgamating factor of the groups—the image they have of their territory is a key element of their vitality. Siegfried is conscious of the diversity of social space: it is he who apprehends this through great monographs of states, as well as through analyses of electoral geography.

These concepts of space have several points in common: they insist on scope much more than on distance; they create a geography of territorial supports much more than they explain political interrelations; and they insist on relations with the environment much more than on ties between the groups.

For some 40 years, progress in human geography has come from a methodical study of the relations being developed between the social and economic factors. Economic geography opened the way: Christaller and Loesch thought out an exceptionally useful tool with the introduction of the notion of a range limit. Political geography has not been able to develop similar instruments during the same period. That explains, even more than the misadventures of geopolitics, the obliteration of this discipline in the last 30 years.

The politicologists are regaining an interest in space while, at the same time, the political sphere attracts a growing number of geographers. That is the result of the introduction of new concepts and methods: The analysis of the flows and relations completes the more traditional ones of the territorial supports. The ideas are derived partly from such systematic analyses as Easton's (1953) and Deutsch's (1953). They have been exploited by geographers such as Jackson (1964) and Soja (1971). Cox (1973) adds to the analysis of the flows of information and methods the one of externality, which introduces, in neighbouring spaces, the domination game. The fashion of perception studies renews the works dealing with the political meaning of space.

Therefore, the theme of centre-periphery introduced itself some years ago in a sphere in full mutation. It is this theme that motivates, to a great extent, the politicologists' interest in space. We should like to locate it in relation to the most important currents of the new political geography.

## SPACE AND THE NEW POLITICAL GEOGRAPHY

The new political geography adds to the study of human and environmental relations—which have made headway largely because of ecological studies—the methodical analysis of all the networks of relations carrying power, authority, and influence. In order to arrive at general interpretations, the attention has been fixed on the elementary mechanism of the political liaisons.

In a first stage, and taking inspiration from Deusch's (1953) works, some authors have tried to measure the political performance of the states against the global intensity of their information flows, in the same way as Soja (1971). The results obtained in countries in the process of formation are interesting, but they are, necessarily, very coarse: They do not allow for an understanding of what is happening within the political system.

Why not try to use in the political field the notion of the limited range, which showed itself to be so fertile in economics? It was applied, and the results are not without interest, but they remain rather trivial: As for the information range, the fundamental threshold separates the oral civilizations from the literate ones. The invention of means of modern telecommunications is of less significance—although it facilitates control, accelerates the dissemination of ideologies, and stops charisma from being a local phenomenon only, always hampered by problems of contact.

The possibility of transmitting messages to faraway places facilitates the formation of networks of specialized political relations—the birth of a nation often goes hand-in-hand with the use of handwriting (for a general view of problems of political ethnology, we refer to Fried, 1967). Consideration of the range limit of political flows throws light on the contrast between oral societies and the others; it justified the social dualism that for a long time in history opposed closed groups, which were gifted with regard to remote relations and local cells still limited to face-to-face exchanges, and where most ties were woven into the neighbourhood.

The means of communication differ not only by the range of the messages, but are also distinguishable by the speed of transmission and the capacity of the lines (Claval, 1978). These variables are of great significance in political analysis: The use of handwriting in a traditional society does not necessarily make it more transparent, for the transmission of news remains slow and the possibilities for dialogue limited; the whole political construction finds itself restrained by these severe conditions. In order to get a good idea, in political analysis, of the importance of the capacity of information transmission, one should classify the power relations with regard to their volume and the speed of the flows they imply.

It is useful to start from the classic distinction between pure power, authority, and influence. First, the relation of pure power is based on the transmission of orders: It seems sparing of communication, but that is an illusion, because those who are subjected to power do not accept it voluntarily; nothing guarantees their obedience. They need supervision: That necessitates, as Foucault (1975) showed so brilliantly, a particular organization of space, i.e., its division into limited areas where supervision can be exercised—that is the principle of all administrative structure. Once the control machine is in place, its functioning leads to the distribution of a considerable mass of information; those who are at the transmission or observation stations can carve for themselves personal influence by withholding some of the news. As long as one does not have speedy and cheap means of communication to institute some parallel networks at one's disposal, and have them control each other, the effective power diminishes with distance: The political systems of pure power are unstable and show a strong centre/periphery opposition; it is a common feature of most traditional states.

Second, the relation with authority limits the quantity of information that must set the political system on its way: It is based on general consent; government decisions are accepted as legitimate and people feel morally obliged to execute them. So, it is sufficient to transmit orders—which is easy—and to keep abreast of what people want—which implies more important flows: Authority is exercised over vast areas without being weakened appreciably by distance.

The real political systems combine power and authority. Not all citizens have the same motivation, and the power game is indispensable to limit deviations. The burden made necessary by the control is eased as the population accepts the authority of the state: Each citizen becomes an observer for the collectivity. The acceptance of legitimate authority is based on the adherence to standards, to social philosophies, and to common ideologies: These are the outlines of areas that share the same political and social values and that limit the range of authority (Gottmann, 1973).

Third, it is therefore the interplay of ideological influences that determines the limits of areas susceptible to submitting themselves to the same authority. In many societies the ideological content of culture changes little—its stability is considered essential: Standards are transmitted from generation to generation through acculturation; the ideological influence is a global phenomenon in which all adults lean on the world of the young that promotes their adoption of the common structures. A specialized body sometimes watches the conformity to the accepted model: Influence stops being diffuse; if the orthodoxy is better conserved in one respect, the ideological influence game models itself from the centre to the periphery and the revealed religions conform with the model.

Ideology is not always considered intangible: Charismatic leaders overturn it in the enthusiasm they arouse; intellectuals transform it in their incessant effort of criticism and revaluation.

The ideological influence game easily takes the form of a domination of the centre over the periphery: The innovation of values is not evenly divided over the various points in space; there, where relations are the most intense and tensions the strongest, the innovations are more numerous; in the image people have of the society in which they live, a gradient is sometimes placed, which accuses the oppositions between centre and periphery. Those who live in prestigious places have more chance to be heard than those who live in the marginal zones. The structure of modern mass communication methods reinforces the polarization effect both within nations and at the international level.

Fourth, besides the ideological influence game, the effects of economic influence should be taken into consideration. These multiply as the exchanges diversify, relationships take on greater distance, and the division of labour is accentuated. The economic influence takes various forms: (1) it shows itself in a positive manner by the opportunity it gives some to innovate, create, and lead; (2) others only possess the power to block, stop, or paralyze. That, however, does not mean that their influence is negligible.

The game of the externalities that create the conglomeration of special services and the focalization of communications and transport networks, the search of scale economies, and the will to procure a market capable of increasing the profit margins turn the polarization of activities into the major feature of the economic geography of the world today: The centre-periphery model explains a big part of the present divisions, but some forces that helped to impose the system at the beginning of the industrial revolution are now breaking down; the telecommunications system limits the advantage of direct access to the sources of information; the de-externalities multiply in the most congested centres; the ease of connections enhances the solidarity of the periphery and allows it to block the proper functioning of the world's economic machinery, if they think fit.

Fifth, the geographers have too often limited the analysis of ecological supports to the study of local relations between man and the environment, but effects of subordination are showing between neighbouring areas: The ecological influence is an important element in the balance of power; the recent works of Cox (1973) remind us of that. The use of the land generates positive and negative externalities in one point; they are not essentially economic, but they are translated in the price of the land and acquire thus a significance in the market economy.

Signs of ecological influence are countless at the neighbourhood level: They are clearly perceptible in the social relations of high density areas and, in order to avoid the nuisance they generate, the owners can only choose between running away and defence by means of zoning. That is how the segregations we frequently find in our cities can be explained.

At the level of large areas, the ecological influence takes another form: It gives some countries, some regions, some crossing points a strategic value. For a long time the military and the specialists of geopolitics have been clinging to these factors of power.

The opposition of the centre and the periphery is one of the elements of these force configurations, but the advantages of the situation deriving from it are generally well balanced: Those who hold a central position fight on interior lines, which increases their maneuvering capacities; those who occupy the fringes are more difficult to reduce to mercy, since their resistance stops only when the totality of the margins of their territory are controlled—unless the interruption of their maritime liaisons puts their forces out of order.

As one sees, the centre-periphery theme is found in several forms in the analysis of the relations of space and power as applied to closed political areas. One also finds it in the study of the open political systems that correspond these days with the international field.

The balance these systems are trying to find can have two origins: It can result from a balance of dissuasion or from a hierarchical order.

(1) The balance of dissuasion is at work in segmentary societies: The social regulation is not the result of any action by any specialized institution; it is born out of fear of a general conflagration. This incites groups to conciliation as soon as some tension appears and, when negotiations break down, the balance is restored by force (Evans-Pritchard, 1937).

Atomic weapons have resuscitated this type of balance in the contemporary world: The Cold War ensures a fragile balance in a world where no nation wields absolute power. The models of a balance of terror do not bring any mechanisms of the centre-periphery type into play—except in special cases, such as the expansionism of societies of the Tiv-Nuer type (Sahlins, 1961).

(2) The most classic model of political balance in the open system is that of

hegemony: A power holds the authority of arbitration and dominates the entire international scene on the condition of not having all of the societies it dominates conspire against it. The model is obviously unstable—the hegemony passes from one nation to another at the end of crises marked by generalized conflict. Such a model is of the centre-periphery type; it implies the effective domination of marginal areas, but also the possibility for these to upset the balance if the supreme nation abuses its power.

The hegemonic model has structured the world since the sixteenth century: First, it was centered in Europe; then it expanded to the entire planet with the birth of the great powers in America and Asia in the nineteenth century.

## CONCLUSION

The present success of the centre-periphery theme somewhat astonishes those who analyze, as we did, the spatial models of power. At the level of political institutions, it is in the traditional societies where the nation starts its structure that the opposition of the centre and the periphery is most marked—it tends to diminish in the developed countries where government succeeds in getting the legality of its actions approved.

At the international level, the hegemonic model prevailed until the first Soviet nuclear explosion. It has not disappeared; each block is still organized on a hierarchical model, but the significance of the nuclear menace is obvious: It puts a ban on the open use of force so as to ensure the policing of the dominated areas; it authorizes, in this way, the peripheral powers to use all their influence to break the hegemony that has been imposed on them. Economic power has shown its strength; world strategy is upset by the problem of energy and raw materials supply: Poles of power multiply.

It is at the level of ideological and economic influences that the centre-periphery theme has the most significance, but here too the situation is changing. The ideological fashions are no longer dictated by the intellectuals of the centre alone; for at least one generation, those of the periphery have seen their audience growing—very often decisively so in the mass media.

In the economic field, the power to innovate, create, and shape the great currents of the life of relations is more concentrated than it has ever been, but there are signs that tend toward a certain devolution. Pollution of industrialized areas, the vigour of union fights, and problems of energy supply provoke the shift of an increasing number of activities toward the periphery—at the same time its blocking power shows up more and more.

The thesis of the centre and periphery owes part of its success to a Marxist current of thought: For those who believe in the ineluctable exploitation of the workers, the increasing concentration of capital, and an indefinite accumulation of profits and property, the image of the centre and the periphery is a good spatial translation of social and economic theory. But the difficulties appear as soon as one tries to define centre and periphery: Space does not enter into the interpretation of the contrasts; one does not set out either its extent or its distance. One explains everything through the domination game, which takes the evanescent and multiple form of the physicists' ether: It constitutes a somewhat mysterious phenomenon where one learns to read, through a socioanalysis (which is incisive on purpose), the Machiavellism of capital.

The study in terms of spatial mechanisms confirms the decisive place of economic facts in the genesis of contemporary oppositions between the centre and the periphery; thanks to the mechanism of polarization, it explains the emergence and the progressive reinforcement of a certain number of world centres. The analysis of the process by which these evolutions take place shows, nevertheless, that the geographical centralism is not a decisive factor for success: On that point the Marxists are right, but they ignore the external influences that are anyhow decisive. Spatial study does not reveal the action of a mysterious dominating power, but the culmination of very concrete political, ideological, and economic effects. Many factors propitious to zones of hoarding of equipment and activities are disappearing. Other balances are in gestation.

The economic field remains the one where the tensions between the developed countries and the Third World are most marked, but the centre-periphery relation shows up the actual process rather badly. The poor countries revolt against the creation of gigantic firms and the impersonalization of decision-making and the relations resulting from it; but the nations of the centre have the same problems: Their peoples are becoming equally wary of the remoteness of authority. The creation of world networks brings with it the delocalizing of the sources of choice; and that is what endangers democratic institutions.

The opposition of the centre and the periphery seems to belong more to the category of expressive images than to that of coherent theories. Many spatial mechanisms of power show a tension between the margins and the heart of the system, but to lead everything back to that dialectic is unrealistic; it risks hiding the deeper causes of the lack of balance in the modern world.

## REFERENCES

CLAVAL, P. (1978) Espace et pouvoir. Paris: P.U.F.

COX, K.R. (1973) Conflict, Power and Politics in the City: A Geographic View. New York: McGraw-Hill.

DEUTSCH, K. W. (1953) Nationalism and Social Communication. New York: John Wiley.
EASTON, D. (1953) The Political System. New York: Knopf.
EVANS-PRITCHARD, E. E. (1937) The Nuer. Oxford: Clarendon.
FOUCAULT, M. (1975) Surveiller et punir. Paris: Gallimard.
FRIED, M. H. (1967) The Evolution of Political Society. New York: Random House.
GOTTMANN, J. (1973) The Significance of Territory. Charlottesville: University of Virginia Press.
JACKSON, W. A. D. (1964) Politics and Geographic Relationships: Readings in the Nature of Political Geography. Englewood Cliffs, NJ: Prentice-Hall.
SAHLINS, M. D. (1961) "Segmentary lineage: an organization of predatory expansion." American Anthropologist 63: 322–345.
SOJA, E. W. (1971) The Political Organization of Space. Resource Paper No. 8. Washington, DC: Association of American Geographers.

# 4

## AMERICA'S CHANGING PLACE IN THE WORLD: FROM "PERIPHERY" TO "CENTRE"?

Alan K. Henrikson

Americans today are not well "located" in international affairs. Having grown accustomed to world leadership since World War II, the modern nonterritorial equivalent of universal empire, they sense they are no longer "the central world power."[1] Many in the United States fear they are being relegated to a periphery—not the periphery of the old colonial "West," but the social and political periphery of a postindustrial "North." Outnumbered, outvoted, and sometimes even outproduced, they feel themselves in a state of incipient historical "decline" or "marginality."

The causes of this apparent descent are not readily understood. Americans are unsure whether their relative loss of status is due simply to the fact that other great powers have risen to challenge their primacy and centrality, and in doing so have "displaced" them, or whether it is due to an upheaval in the basis of the international system itself, in the underlying hierarchical-locational structure of international relations.[2]

President Carter, speaking at Notre Dame University on May 22, 1977, addressed this concern of his countrymen explicitly. "By the measure of history," he said, "our nation's 200 years are brief; and our rise to world eminence is briefer still. It dates from 1945, when Europe and the old international order both lay in ruins. Before then, America was largely on the periphery of world affairs. Since then, we have inescapably been at the center."

But perhaps no longer. "Historical trends"—the relaxation of the unifying threat of conflict with the Soviet Union, the moral crisis over the Vietnam War, the economic strains of the 1970s and the apparent inability of industrial democracy to provide sustained well-being, and, above all, the passing of colonialism with the resultant "new sense of national identity" in nearly 100 new countries—have weakened the foundations of the system in which the United States once predominated. It is, President Carter declared, "a new world," which calls for "a new American foreign policy."

Americans should no longer expect, Mr. Carter warned, that the other 150 countries will "follow the dictates of the powerful." The era of "almost exclusive alliance" with the Atlantic states is past, and "a wider framework of international cooperation" is needed. "We will cooperate more closely," he affirmed, "with the newly influential countries in Latin America, Africa and Asia. We need their friendship and cooperation in a common effort *as the structure of world power changes*" (emphasis added; Carter, 1977).

President Carter's use of the geometrical terms "center" and "periphery" in discussing this changing structure, and America's place in it, draws attention to a dimension of our theme that is nowadays often overlooked or minimized: the spatial or geographical. To see the history of American foreign relations in this way is, however, natural. The various changes that have occurred in America's "place in the world" probably add up to a greater actual and imagined movement in space than that experienced by any other country, at least in so brief a time.

How does the "center-periphery" schema help us to understand the shifts in America's world position? The answer, it seems to me, lies in the relationship of polarity that is inherent in the center-periphery model. Let me briefly explain.

A center-periphery system is characterized by a certain physical, social, and psychical distance between an interrelated core and margin (Eisenstadt, 1977:72).[3] If a center-periphery structure is to endure, core and margin must be kept distinct. If essential elements of the center migrate to the periphery, or if substantial parts of the periphery are absorbed at the center, the relationship between them may break down. Similarly, if the center "sinks" or the periphery "rises," a breakdown in the relationship can occur. If either of these kinds of movement, horizontal or vertical, becomes massive enough—that is, if a periphery heavily engulfs a center or decisively surpasses it in some respect—then there can occur an actual reversal of roles: a switch in polarity.

Although the historical processes that underlie such crossovers are usually incremental and long term, the role reversals themselves may appear to be sudden. The explanation of this felt abruptness may in some cases be mainly psychological. Like the instantaneous image shifts in the experiments of Gestalt psychologists, center/periphery reversals in geopolitical fields may also seem to occur rapidly, especially when accompanied by dramatic events.

The structure of a fluid situation may require a new idea or theme in order to become crystallized. In international affairs this is especially the case, since the participants are so varied, their relationships so numerous, and their environments so vast. Often the best representation of an evolving international system is a map, a visual analogue to political and other facts in their geographic setting.[4] Some use will be made of maps, both imaginary spatial images and actual cartographic representations, in the present essay. Maps can have iconic value. They represent hierarchies and locations symbolically. As evidence and as emblem, they mould as well as mirror historical circumstances.

At a time of crisis, the mind may oscillate between two "views," or impressionistic interpretations, of the course of world events. If supported by a theory of international relations, that is, if a *perceptual* change is accompanied by a *conceptual* change, a sensed shift in polarity may become mentally fixed. A new "paradigm," in Kuhn's (1970) sense of that term, controls the future direction of thought.

The classic case of a lasting center/periphery reversal, of peculiar analogic appropriateness to our present theme, is the sixteenth-century shift from a Ptolemaic, earth-centered concept of the universe to the Copernican, sun-centered world view. Once the new logic of heliocentrism was understood, the old "normal science" of geocentrism became not simply outmoded but nonsensical.[5] Similarly, if a new scientific theory is offered and accepted as a rationalization of a geopolitical role change, it may make the change intellectually (if not practically) irreversible.

The international history of the United States contains a number of such perceptual-conceptual shifts, most but not all of them related to shifts in the world position of Europe. The greatest of these was that to which President Carter referred—the great globewide center/periphery reversal of the 1940s.

A shift from periphery to center, one would think, assuming the irreversibility of such historical movements, could occur only once. In fact, America and Americans have arrived at, or acquired, a position of "centrality" a number of times. The apparent illogic of this is explained by the realization that America's centrality has been gained in different—increasingly larger and more complex—contexts.

Originally, as President Carter stated, America was "largely on the periphery." By this he presumably meant the periphery of world affairs as dominated by Europe. The idea is rooted in America's colonial past. For nearly 200 years the lives of British settlers in North America were regulated by the mercantilist policies of distant London.

The notion of American peripherality also derives, however, from ancient geographical tradition. For Eratosthenes, and most of his successors down through the Middle Ages, the earth had but three parts—Europe, Asia, and Africa—rimmed by a narrow, circumfluent ocean. The discovery by Columbus of the "Indies" (as he believed them to be) did not immediately destroy this inherited picture of earth. A new continent would be a "fourth part" of the world—a conceptual impossibility. Moreover, with six-sevenths of the *orbis* land, there was hardly any room for a new part! Until Martin Waldseemueller's famous 1507 map, which gave America its label, the new land was conceived, and cartographically represented, as the easternmost extension of Asia (James with James, 1972; Thrower, 1972; Boorstin,1976. When the full expanse of the globe was finally appreciated, the American continents were often placed on the left side of joined-hemisphere maps—a solution that, while it kept Eurasia at the center of its orbit, made America central as well (Figure 1). The anomalous fourth part of the Old World became the core of a New World—a Great Satellite.

The first major step in the actual fulfillment of this schematic "Western Hemisphere" centrality was, of course, the American Revolution. The Declaration of Independence, a document of not only political but geographical

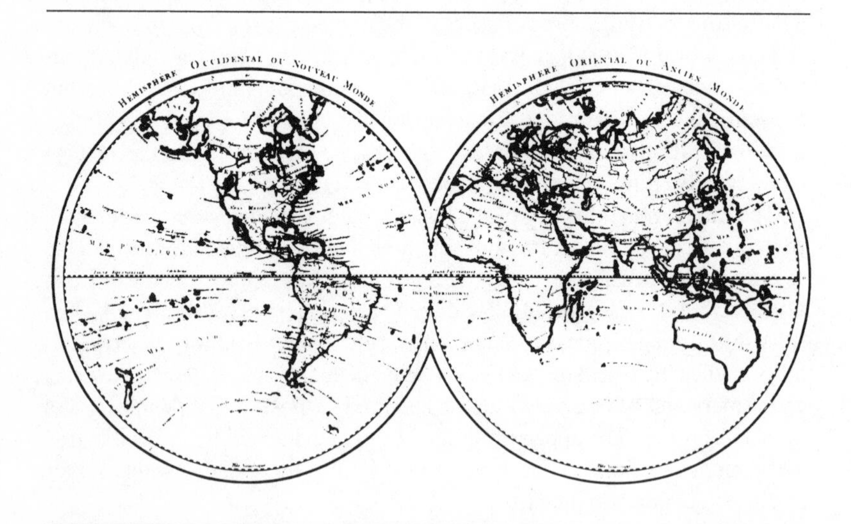

FIGURE 4.1: Mappemonde en deux hémisphères
SOURCE: Harvard Map Collection, Harvard College Library

import, proclaimed the need for Americans "to dissolve the political bands, which have connected them with another, and to assume among the powers of the earth, the separate and equal station to which the Laws of Nature and of Nature's God entitle them." Thomas Paine in *Common Sense* (1776) provided a Copernican-Newtonian justification for America's geopolitical independence. "In no instance," he argued, "hath nature made the satellite larger than its primary planet, and as England and America, with respect to each other, reverses the common order of nature, it is evident they belong to different systems: England to Europe, America to itself" (Kramnick, 1976: 91).[6]

A quaint yet characteristic expression of this primitive America-centrism was the late eighteenth- and early nineteenth-century cartographic practice of placing the prime meridian of longitude within the American orbit. Some of the first maps of the United States show the central meridian running through Philadelphia. The "Meridian of Philadelphia" gradually gave place to a Washington, D.C., zero meridian. On L'Enfant's plan of Washington, the 0°0′ line passes through "Congress House." Other contemporary *projets* put the line through the "President's House" (Pratt, 1942).[7]

This use of locally defined zero meridians was, of course, partly just cartographic convenience—the tradition, familiar in Europe, of putting the 0°0′ line through the largest, most important city on the map. In the Philadelphia and Washington cases, however, it was a reflection of something more: an expression of an intention to lay down an entirely new geographical frame of reference. The Meridian of Greenwich remained a symbol of monarchism, a vestige of colonial subordination. "Are we *truly* independent, or do we appear so," a patriot asked in 1819, when on leaving his own country every American is "under the necessity of casting his 'mind's eye' across the Atlantic, and asking of England his relative position"? This was calculated to "wean the affections. . . . What American seaman has not experienced this moral effect?" (Stanton, 1975:7).

A republican pole—or an alternative imperial pole—was needed. This position was assumed by Washington, the District of Columbia. Some of its builders ambitiously conceived of it as the "Rome of the New World." (Given the fact that most of South America lies longitudinally to the east of the United States, a Washington zero meridian is actually not a bad choice for the vertical axis of the New World.)

What was the character of this new American Empire? Initially it was defined only negatively, in contradistinction to Europe. It was conceptually a void—a counterculture, a counterpolity, a countereconomy, even a countergeography. The comparative underdevelopment of the American idea is evident in Washington's Farewell Address. "Europe has a set of primary interests, which to us have none, or a very remote relation," it stated. "Our detached and distant situation invites and enables us to pursue a different course" (Gilbert, 1961:144–147). What this course might be Washington did not specify.

President Jefferson had a more positive, idealized, broadly pan-American conception. "America has a hemisphere to itself," he wrote to Alexander von Humboldt, who had explored Latin America. "It must have a separate system of interest which must not be subordinated to those of Europe" (Whitaker, 1954:29). Henry Clay had an idea about how the American orbit might be internally organized. "It is within our power," he declared, "to create a system of which we shall be the center and in which all of South America will act with us" (Rippy, 1958:4–5).

The Monroe Doctrine may be taken as the point at which, in American and in certain European eyes, the American system began to have parity with the European system. It seemed to be gaining gravitational equivalence. Such was the increase in America's overall mass that Jefferson, asked by Monroe whether to accept the British Foreign Secretary's proposal of a joint declaration on Latin American independence, imagined that, by acceding to Canning's proposition, the United States might "detach her from the bands, bring her mighty weight into the scale of free government, and emancipate a continent at one stroke" (Bartlett, 1970:174–175).

John Quincy Adams, defending his administration's participation in an inter-American congress at Panama in 1826, argued that the trebling of the population, wealth, and territory of the United States since Washington's Farewell Address made possible a reversal of the founding father's formula. Rather than reemphasizing that Europe had a set of primary interests with which the United States must not interfere, Adams, in a message to the House of Representatives on March 15, 1826, stressed just the opposite: The time had arrived when "*America* has a set of primary interests which have none or a remote relation to Europe" (LaFeber, 1965: 136). America's independence, or polar position, was thus complete.

Westward expansion consolidated this image of hemispheric independence and unity. By quick, giant steps—the Louisiana bargain with France (1803), the Transcontinental Treaty with Spain (1819), the Oregon Treaty with Great Britain (1846), and the Treaty of Guadalupe Hidalgo with Mexico (1848)—the United States extended its domain across North America. In so doing, it defeated any European thought of maintaining a North American "balance of power" and any Latin American thought of offering direct rivalry.

The United States now "faced" both the Atlantic and the Pacific—a fundamental change in geographical outlook and focus. To some Americans, such as Senator Thomas Hart Benton, the country's intermediate situation between Europe and Asia presented it with an opportunity to become the prime route to the Orient, thus fulfilling the original dream of Columbus. "The European merchant, as well as the American," Benton prophesied in 1849, "will fly across our continent on a straight line to China. The rich commerce of Asia will flow through our centre" (Smith, 1950:28–29). His home town of St. Louis, he hoped, would become the emporium of this global East-West trade.

This idea of mid-America as the next great center of international exchange was supported by a deterministic geographical theory derived from Humboldt's delineation of global equal-temperature zones: that the zone of (northern) temperate climates is also the path of world progress. Benton's associate, the speculator-soldier-politician William Gilpin, envisioned an "Isothermal Zodiac"—a great undulating common-temperature belt, about 30° wide, girdling the northern hemisphere between roughly the twenty-fifth and fifty-fifth parallels and, happily, including all of the territory of the United States. Through it ran an "axis of intensity," at about the fortieth parallel (St. Louis: 39° N), which had a mean annual temperature of 52°. "Within this isothermal belt, and restricted to it," wrote Gilpin, "the column of the human family, with whom abides the sacred and inspired fire of civilization, accompanying the sun, has marched from east to west, since the birth of time." Upon it had been constructed "the great primary cities"—the Chinese, Indian, Persian, Grecian, Roman, Spanish, and British—and now "the republican empire of the people of North America."

Topography and hydrography, too, favored the world centrality of North America. The Great Basin of the Mississippi was to Gilpin "the amphitheatre of the world." In contrast to the interiors of the other continents, that of North America presented toward heaven "an expanded bowl," catching and fusing whatever entered within its rim. Other continents presented "a bowl reversed," scattering everything into radiant distraction. "In geography the antithesis of the old world," he judged, "in society we are and will be the reverse"(Boorstin, 1965:233).[8]

Given such grandiose views, it is not surprising that the United States began around 1850 to appear at the center of world maps (Figure 2). Previously, on most American as well as European global maps, the United States had been off to the left, in the West. Following the U.S. victory over Mexico, the discovery of gold in California, the triumph of the Union in the Civil War, and the purchase of Alaska, the country seemed to many American map makers to deserve a more central position.[9]

The principal reason for this shift of global frame was, one suspects, not the increase in power, wealth, and solidity of the young Uncle Sam, but the

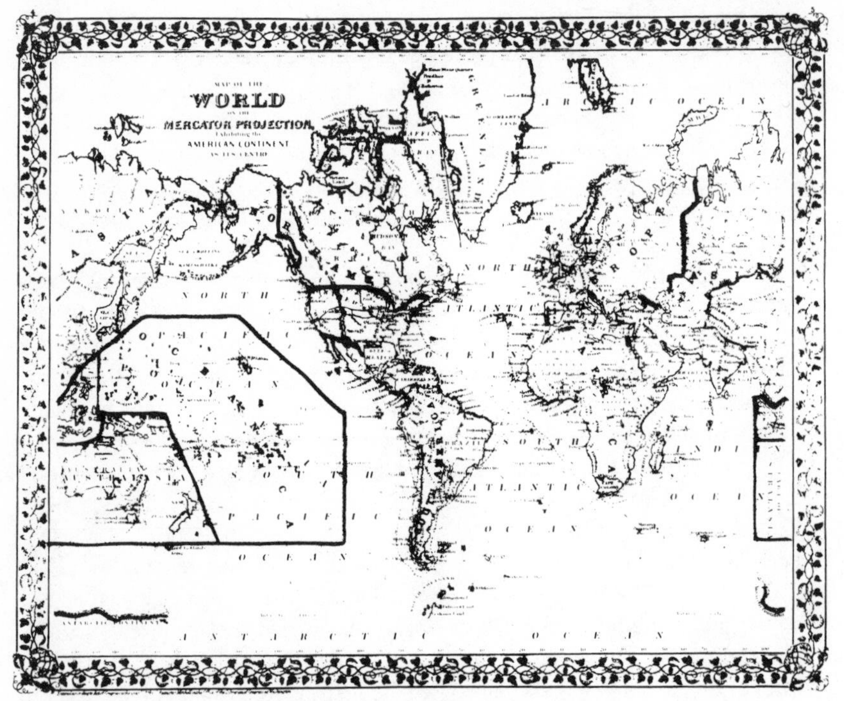


FIGURE 4.2: American world map with the United States at the center
SOURCE: From S. Augustus Mitchell, *New General Atlas*

change in his geographical stance—that is, the fact that he now stood astride a continent, one foot on the Atlantic and the other on the Pacific.[10] It only "made sense" for the self-proclaimed broker betwen Occident and Orient to be in the middle of the picture.

Despite this spread-eagle imagery, the United States was, in a political and military sense, isolated and marginal in world affairs throughout most of the nineteenth century. Its position and perspective were still essentially "continental." Captain Alfred Thayer Mahan (1890:42) described this self-contained outlook well in his book, *The Influence of Sea Power upon History:*

> Except Alaska, the United States has no outlying possession—no foot of ground inaccessible by land. Its contour is such as to present few points specially weak from their saliency, and all important parts of the frontiers can be readily attained—cheaply by water, rapidly by rail. The weakest frontier, the Pacific, is far removed from the most dangerous of possible enemies. The internal resources are boundless as compared with present needs; we can live

off ourselves indefinitely in 'our little corner,' to use the expression of a French officer to the author.

The Spanish-American War broke this defensive, continentalist shell. For the first time in its history, the United States was connected geopolitically with the wider world. In the conquered Philippines, the onetime distant periphery gained its own distant periphery. Partly because of this possession, the American Republic was increasingly described as a "world power" (Coolidge, 1912; May, 1961). In truth, it was only a Caribbean power. Nonetheless, its Philippine commitment marked a major location shift.[11]

From the prospect of Europe, America was still remote—in some ways *more* remote than it had been. Halford Mackinder (Pearce, 1962: 67, 261, 262) in his famous 1904 lecture on "The Geographical Pivot of History," declared: "The United States has recently become an *eastern* power, affecting the European balance not directly, but through Russia, and she will construct the Panama Canal to make her Mississippi and Atlantic resources available in the Pacific. From this point of view the real divide between east and west is to be found in the Atlantic ocean" (emphasis added). Mackinder's illustrative maps placed the American "satellite," as he described it, on the "outer or insular crescent"—on the right, or eastern, side of the map as well as on the left, or western, side.[12] If anything, America, pivoting elliptically around the Eurasian "Heartland," was thus more "peripheral" than it had been.

This impression was illusory. The very idea of marginality suggested mobility, and the offsetting of continental power by maritime power. Dewey's victory at Manila Bay and the round-the-world cruise of the Great White Fleet announced America's presence on the high seas.

In closely related developments, the diplomatic role of the United States also soon expanded. The participation of the United States in the Hague Peace Conferences, its singular part in ending the Russo-Japanese War, and it mediatory diplomacy during the first Moroccan crisis all bespoke a new and wider American centrality in world affairs. No continent was completely irrelevant to the United States any longer.

Theodore Roosevelt, probably the first American president to have a real sense of a *world* balance of power, saw heroic possibilities for the nation. In 1910 he told Baron von Eckardstein, former German ambassador in London, that, should Great Britian fail in her traditional role, "the United States would be obliged to step in at least temporarily, in order to restore the balance of power in Europe, never mind against which country or group of countries our efforts may have to be directed." "In fact," he judged, "we ourselves are becoming, owing to our strength and geographical situation, more and more the balance of power of the whole world" (Beale, 1956: 447).

Woodrow Wilson forsook "world balancing" for "world leadership." If Roosevelt's typical image had been that of a global scale. Wilson's was that of a global platform. He simply thought on a different plane—that of ideology. On this level, his horizons were unlimited. In his second inaugural address (March 15, 1917), he declared that "we realize that the greatest things that remain to be done must be done with the whole world for stage and in cooperation with the wide and universal forces of mankind, and we are making our spirits ready for those things. We are provincials no longer" (Inaugural Addresses, 1961: 203–206). He said this even before the United States had entered World War I. Afterward, America's global destiny was even clearer to him: "There can be no question of our ceasing to be a world power. The only question is whether we can refuse the moral leadership that is offered us, whether we shall accept or reject the confidence of the world" (Weinberg, 1935: 470).

Wilson failed in his aim partly because he tried too hard to be cosmopolitan, to transcend the barriers of physical and cultural distance. In attempting to bring the United States into a world organization, he lost his sense of America's "place," and of his own base of political support. This was a "geopolitical" misjudgment far greater than his particular errors in, for example, defining the Italian-Austrian boundary or assigning "Class C" mandates. It is conceivable that had he possessed a better appreciation of the psychological gulfs between continents and, accordingly, proposed a more "regionalized" League of Nations, his experiment might have turned out more successfully.

World War I and Wilson's universalist leadership did, however, greatly expand Americans' world knowledge and interest. This was most impressively manifested in the economic realm, where an important reversal in polarity occurred. The United States, throughout its previous history a net borrower, became a net lender. "Since the beginning of the World War," Bowman noted (1928: iii), "the United States has increased its foreign investments fourfold, doubled its foreign commerce, and become the creditor of sixteen European nations." The geographical scope of its stake was second to none. "If our territorial holdings are not so widely distributed as those of Great Britain, our total economic power and commercial relations are no less extensive."[13]

New York emerged as a world financial center, rivaling London. The extent of the world's dependence on the American capital market did not become fully apparent, however, until after Black Thursday on Wall Street. As Braudel has reflected (Time, 1977), "In the world of exchange, there's always a central zone, an intermediary zone, and a peripheral zone. In 1929, the so-called Dark Year, the center of the world, which was London, passed to New York, peacefully."

World War II completed the shift of the United States from periphery to center on other levels. A group of distinguished European refugee and American intellectuals, stunned by the fall of France, issued a manifesto (Agar, 1940), *The City of Man: A Declaration on World Democracy,* in which they lamented that Europe's "solar period" was over. The new sun for mankind would have to be America—whose light, they urged, must shine more steadily and universally.[14]

This meant nothing less than a transatlantic "Copernican Revolution," a reversal in the direction of spiritual (if not yet cultural) gravitation and radiation. The dawn of the new U.S.-dominated age was proclaimed by publisher Henry R. Luce in his influential tract, *The American Century* (1941: 39). Objecting to former President Herbert Hoover's observation that America was "fast becoming the sanctuary of the ideals of civilization," he retorted that it now "becomes our time to be the powerhouse from which the ideals spread throughout the world."

There was a reversal in America's world geostrategic situation as well. Before Pearl Harbor, the major military doctrine of the United States was "hemisphere defense." After Pearl Harbor, it became "global offense" (Figure 3).[15] Either encircle or be encircled, President Franklin D. Roosevelt warned in a fireside chat of February 23, 1942. "We must all understand and face the hard fact that our job now is to fight at distances which extend all the way around the globe" (Rosenman, 1950). Otherwise, the United States would become detached from its allies and end up fighting the Axis alone.

From the American perspective and, to a lesser extent, from the perspective of other nations, World War II was, in an unprecedentedly literal sense, a global war. The relationship of the United States to the major theaters of battle was such that a new picture of the world—a new global strategic map—was needed. Cylindrical map projections, such as the conventional equator-based Mercator, failed to show the continuity, unity, and organization of the "worldwide arena," as Roosevelt called it. Hence, other map projections came into fashion, notably the North Pole-centered azimuthal projection. The masterpiece of this type was probably Richard Edes Harrison's "One World, One War" (Figure 4), appearing originally in *Fortune* magazine and widely reproduced for military training and other purposes.[16]

The position of the United States on these polar maps was usually a central one—at or just below the geometrical center (usually 90° N) and on the vertical axis (usually 90°W–90° E). This was, perceptually speaking, the prime location. As Arnheim (1977) points out: "Any picture space is dominated by the vertical central spine, which is experienced as pointing

THE STRATEGY OF SECURITY

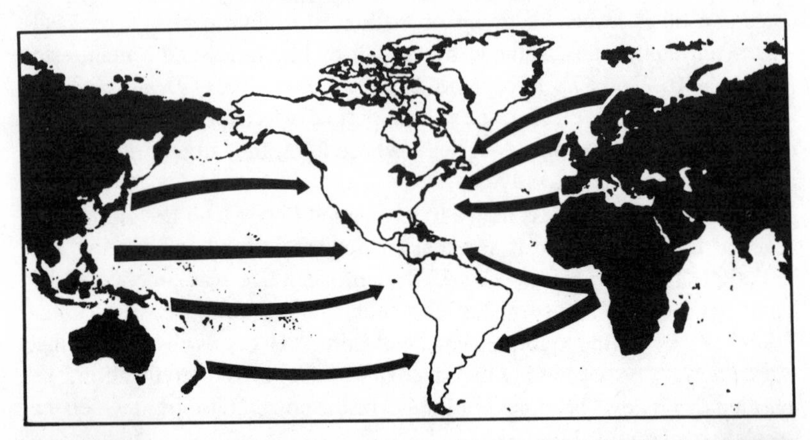

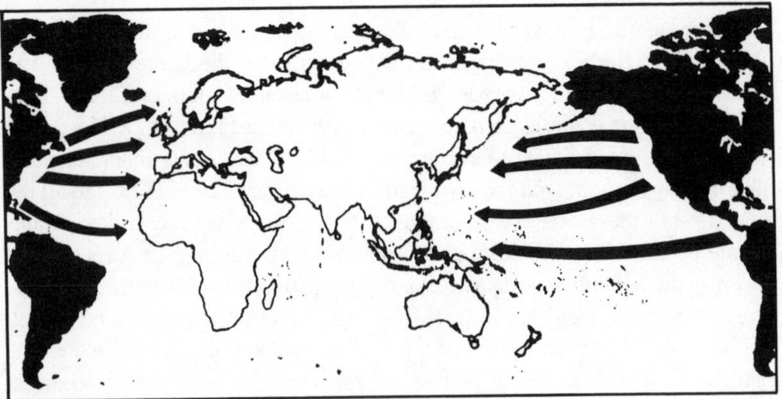

THE FUTURE OF THE WESTERN HEMISPHERE

FIGURE 4.3: From Western Hemisphere defense to global offense
SOURCE: Map by J. McA. Smiley from *The Geography of the Peace* by Nicholas John Spykman, reproduced by permission of Harcourt Brace Jovanovich, Inc.

upward—probably in conformity to the forward direction of the viewer. When I look at a picture, the upward direction in the picture corresponds to where my nose points and where I go if I move. The lateral directions are symmetrical to each other and constitute a (less dynamic) base."[17]

The new transpolar focus did not come naturally to Americans. Although

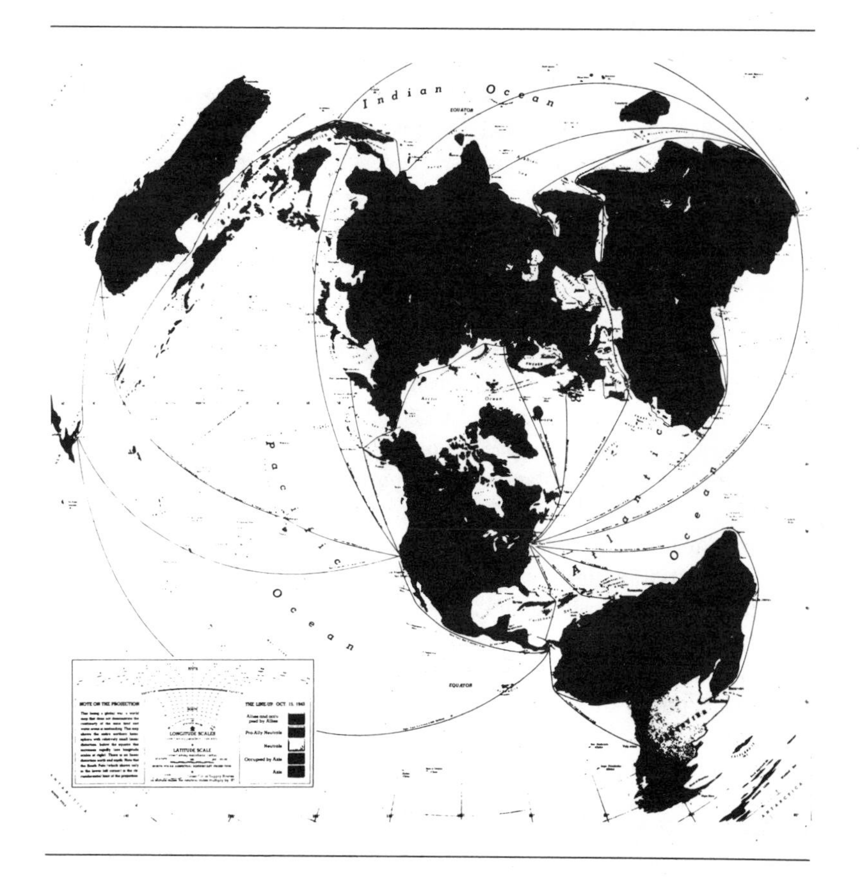


FIGURE 4.4: One world, one war
SOURCE: Richard Edes Harrison for *Fortune Magazine.* © 1942 by Time,Inc.

their maps had conventionally been North-"oriented," the course of their history had largely run eastward and westward. Except for a few Arctic enthusiasts, such as the explorer-scientist Vilhjalmur Stefansson, the Northland was frozen, inert, and Canadian. Alaska, in most minds, was still "Seward's Folly." In order to make the North the scene of potential activity—to make it "dynamic"—cartographers drew arrows to suggest cross-polar movement.[18]

An important consequence of viewing these wartime North Pole-centered maps was a new awareness of the proximity of North America to Eurasia, and vice versa. The Arctic Ocean became, psychologically, an inland sea—a circumpolar "Mediterranean." Russia, previously thought of as being on the opposite side of the earth, suddenly appeared overhead. Russia had been situated "behind" the countries of western Europe and eastern Asia; now it was located directly "in front of" them. No longer, as Lippmann (1943: 145) ominously warned, would American-Russian relations be controlled by "the historic fact that each is for the other a potential friend in the rear of its potential enemies." What factors, then, *would* control relations between the two powers that Tocqueville had identified as marked out by the will of heaven to sway the destinies of half the globe?

The old "Heartland" theory, according to which the United States was a mere "satellite" in orbit around the planetary "World Island," was clearly no longer tenable. The United States was now too independent, too strong, and too far-reaching in its influence to be considered only a peripheral power. Columbia University geographer George Renner (1944: 44,47), fascinated by the technology of polar aviation, proposed an adjustment: that the Heartland be expanded and shifted upward to include the interior parts of all the land masses ringing the Arctic Mediterranean. Mackinder (1943: 598,602) himself adjusted his concept by encompassing the Heartland and "Midland Ocean" (North Atlantic-Arctic) with a new "great feature of global geography: a girdle, as it were, hung around the north polar regions" (Figure 5). Spykman (1944: 40–41) further detracted from the Heartland's centrality by focusing on the critical "Rimland" (Mackinder's "Inner or Marginal Crescent"). To this vital intermediate zone between land power and sea power the United States, like most other great powers, had direct access. Cressey (1945: 245–246) completed the displacement of the old world core. "If there is anywhere a world citadel or Heartland," he proposed, "it may well lie in North America rather than in Eurasia." He justified his conclusion: "The American continent has adequate size, compact shape, internal accessibility, a central location, good boundaries, access to two oceans, favorable topography, rich minerals, excellent climate, and a dynamic spirit of its people."[19]

At war's end, the United States thus emerged as the geostrategic center of the world, at least the geographic center of global strategic decisions. Political lines followed military lines. Washington, D.C., was finally the "Rome" to which all roads—not just those of the New World—seemed to lead. Americans began to think of themselves, as Woodrow Wilson had urged, not merely as U.S. nationals but as citizens of the world. Others, too, began to look upon the United States as world capital—an image strengthened by the selection of New York as the site of the United Nations Organization (Eichelberger, 1977).

Europeans, particularly, had a sense of the reconstellation of world

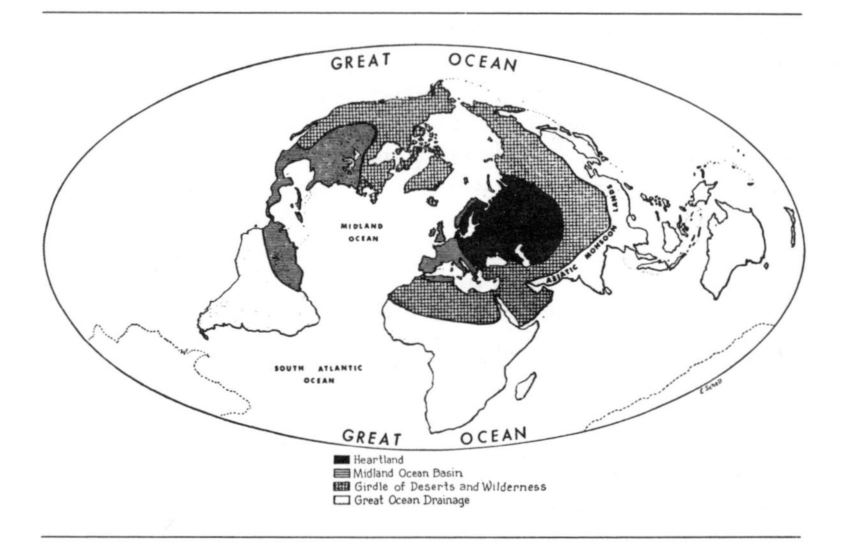


FIGURE 4.5: Mackinder's world (1943)
SOURCE: From *Geography and Politics in a World Divided* by Saul Bernard Cohen. © 1963 by Random House, Inc.

forces. They felt themselves in a position similar to that of the Greeks in 146 B.C.: America's power center had coalesced and Europe's had collapsed. Luce's *The American Century* had its complement in Fischer's *The Passing of the European Age* (1948) and similarly titled works (Weber, 1948; Holborn, 1963). "The capital fact," André Malraux reflected,

> is the death of Europe. When I was twenty years old [in 1921] the United States was approximately in the position of Japan today in terms of world importance. Europe was at the heart of things and the great superpower was the British empire. But now all dominating forces in today's world are foreign to Europe. The great power in the world is the United States and, to the side, there is the Soviet Union. Europe has virtually disappeared as a factor and it took astonishingly little time for this change to come about. Two centuries ago the United States was not even a nation; now it is a colossus [Sulzberger, 1973: 4-5].

The American colossus, which hitherto had towered only over the Western Hemisphere, now had a foot in Europe and another in Asia. By subtle degrees, its overseas wartime missions and occupation responsibilities were transformed into alliance commitments—the Rio Pact, NATO, ANZUS Pact, Philippine Treaty, and so on. Washington truly held the reins of a global politico-military coalition.[20]

This situation was historically unprecedented. "The radical novelty of our present position," wrote Lippmann (1952: 31–33), "is, as seen with American eyes, that we have now become a principal power." Whereas during World War I, and even part of World War II, the United States had been "an auxiliary power, a supporting and reinforcing power, called in to redress the balance of the Old World," it now had the central position and responsibility. Many of its former allies, and even its enemies, were "auxiliaries" in helping it maintain the balance of the world as a whole.

The main strategy adopted by the United States for safeguarding the global balance was that of "containment." As originally defined by Kennan (1947), this strategy was "the adroit and vigilant application of counter-force at a series of constantly shifting geographical and political points, corresponding to the shifts and manoeuvres of Soviet policy." To meet expected Soviet aggression, the inner and outer defense perimeters of the American sphere were fanned out in all directions.[21]

The flaw in the containment strategy, as Lippmann (1947: 21) and other critics pointed out, was that it relied on the "feeble or disorderly nations, tribes and factions around the perimeter of the Soviet Union" rather than on the compact nucleus of the Atlantic Community—the center. Moreover, the strategy seemed negative, reactive rather than affirmative. The adversary seemed to have the initiative. To follow such a policy could only debilitate.[22]

There was a further, psychological danger in the containment policy: by focusing on the outer margin of the Soviet bloc rather than on the Western sphere, American officials might shift the "center" to Moscow, leaving the West "centerless." On a contemporary map of the American perimeter of defense around the communist core, the larger American sphere can easily become recessive and the smaller but more compact Soviet-Chinese sphere dominant (Figure 6).[23] This perceptual switch has its counterpart in the ideological realm: The "Free World" becomes a mere anti-"Slave World," much as "America" had once been a mere anti-"Europe."

With the 1955 Bandung Conference, the central position of the United States in world politics began to erode in another way. New political centers in the "South"—Djakarta, New Delhi, Cairo, Lagos, Accra, Brasilia—emerged. A "Third World" rose to challenge the Old and the New Worlds. This southward shift in a hitherto largely "northern" international system was supported by global demographic trends.

So fundamental were these changes that a new *Weltbild* was needed. The German historian Arno Peters produced one cartographically. Noting that on a conventional Mercator-projection map the central horizontal axis is well above the equator, a placement that has the effect of exaggerating the size of North America and Eurasia and dwarfing Latin America and

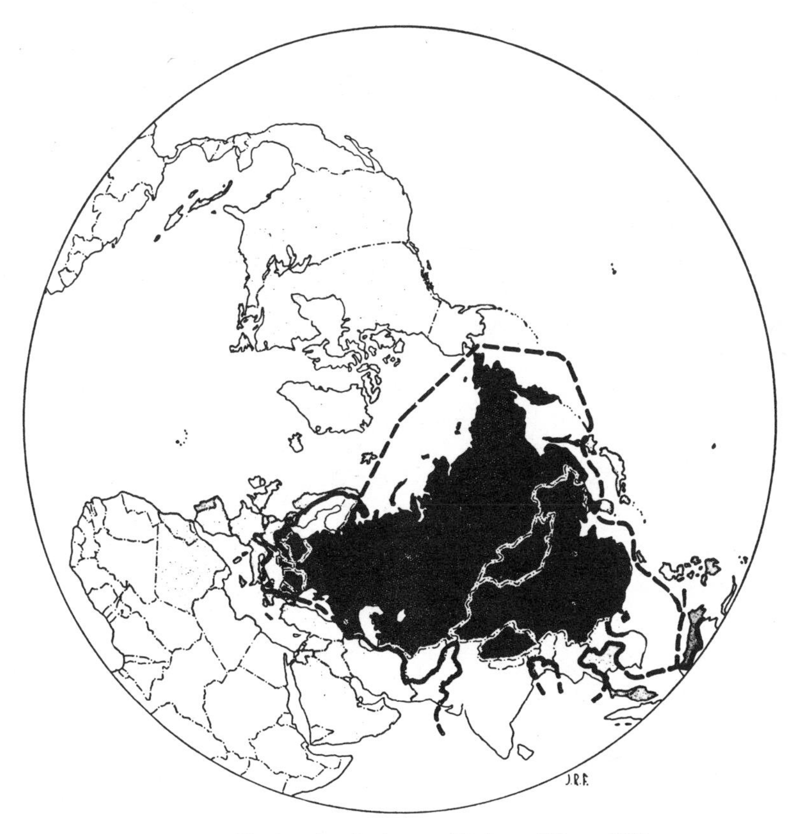


The American Perimeter of Defense: Winter, 1955.

FIGURE 4.6: The American perimeter of defense (1955)
SOURCE: From *Principles of Political Geography* by Hans W. Weigert et al. © 1957 by Appleton-Century-Crofts, Inc.

Africa, he recommended an equal-area projection, centered on the equator, in which this "Northern Hemisphere" bias is rectified (Figure 7).[24] This shift in cartographic emphasis, while directed mainly against "Europe-centrism," also counteracts "America-centrism." The United States is shunted into an upper corner, a "northern" periphery.

There was as yet no gainsaying, however, the continued world centrality of the United States in economic and military affairs. This power, during the 1950s often latent, became overt in the 1956 Suez crisis and, later, the 1962 Cuban missile crisis. Britain and France could not sustain their invasion of

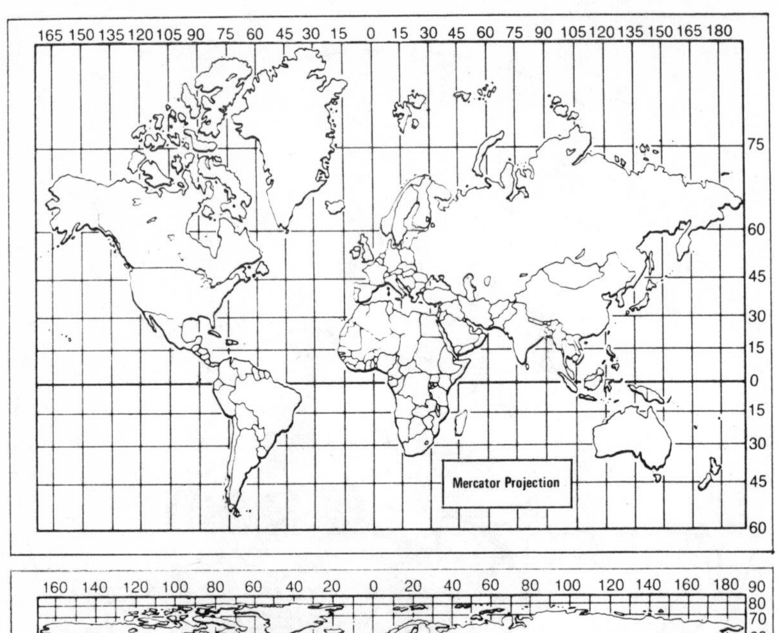


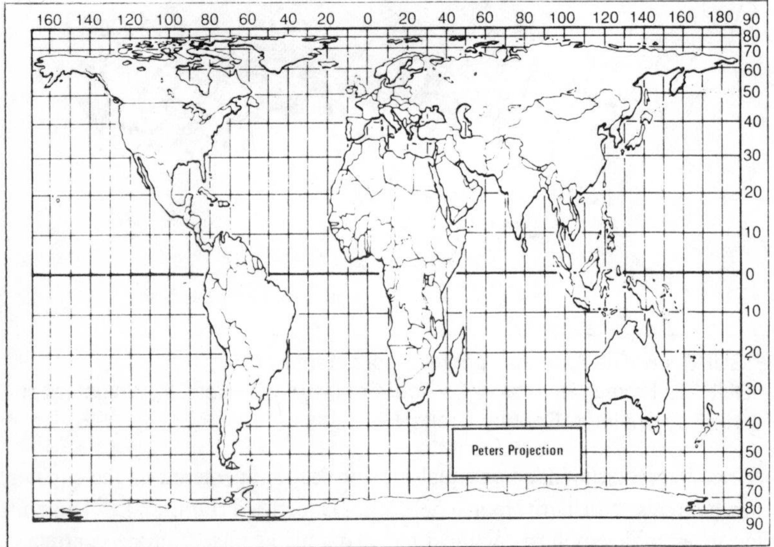


FIGURE 4.7: "Northern" world and "Southern" world
SOURCE: Los Angeles *Times* map by Don Clement © 1973. Reprinted by permission

Egypt without the concurrence of Washington, upon whom they were dependent for oil, credits, and even confidence. During the Cuban crisis, the

Soviet Union could not, in the end, complete its installation of nuclear missiles near the United States. The reasons for these failures were the same: the acknowledged role of the United States as the primary upholder of order in the world.

In the 1960s it was common to refer to the United States as "the only truly global power." The perverse proof of this was the war in Vietnam—America's "first imperial war," as Liska (1967) characterizes it.[25] This conflict was waged over a distance of 8000 miles, with seemingly inexhaustible reserves, ostensible moral conviction, and apparent political impunity. No military objective seemed impossible. Wohlstetter (1968), rejecting the assumption that a nation's power and knowledge must fall off steeply with distance from home, argued that, to the contrary, the advancing technology of transport and communication made even the most faraway country accessible—and, by implication, conquerable.

The American antiwar reaction, which grew in the aftermath of the 1968 Tet offensive, led to a full-blown reexamination of America's historic place in the world. One result was a rediscovery of what Ball (1968: 351–352) has termed "the difference between center and periphery." Americans had been navigating, Ball suggested, "by a distorted chart—like something drawn by a medieval cartographer, in which Vietnam appears as a major continent lying just off our shores and threatening our national existence."[26] He urged that it be put into true scale and back into its proper Southeast Asian setting again.

The new foreign policy search, as Ball (1968: 353, 356–357) put it, needed to be for "a solid base" on which some durable system of power could be erected. For him, as for like-minded members of the subsequently organized Trilateral Commission, this was the "nuclear," or central, relationship with the other nations of the Atlantic Community, plus Japan. In particular, he favored a unified Western Europe, which could assume with the United States "a political partnership that fully reflects the common interests of a common civilization."

Other theorists considered such an alliance too narrow, too insufficient. The United States, Henry Kissinger indelicately pointed out, had global interests, whereas Europe had only regional interests. The best way for American diplomacy to resolve the central issues, especially the strategic arms race, and even to handle such peripheral matters as Vietnam, was to deal directly with the Soviet Union and the People's Republic of China. Hence the diplomacy of Détente and the unprecedented presidential trips to Peking and Moscow.

The resulting fluidity of international relations was superficially rationalized by the concept of a "pentagonal" world in which all great powers—

allies or adversaries—functioned more or less alike. President Nixon explained: "I think it will be a safer world and a better world if we have a strong, healthy United States, Europe, Soviet Union, China, and Japan, each balancing the other, not playing one against the other, an even balance" (Time, 1972: 14). Implicitly, this was an admission of the descent of the United States from superiority to parity, from being *the* center to being *a* center.

The 1973 October War and Arab oil embargo further detracted from an American sense of world centrality. The Nixon Administration's attempt to organize a "multilateral" response to Arab economic blackmail failed. Europe, which imported more than 80 percent of its petroleum supplies from the Middle East, could not afford the likely penalty of such a broad confrontation. A map of world oil movements (Figure 8), showing Europe's much greater vulnerability, makes the logic of Europeans' refusal to follow American leadership plain.[27] Given this divergence of interest, it began to seem once again, as it had for Mackinder in 1904, that "the real divide" between East and West lay in the Atlantic. The emerging rift was overcome, however, as America's imports of Middle Eastern oil also rose to huge amounts.

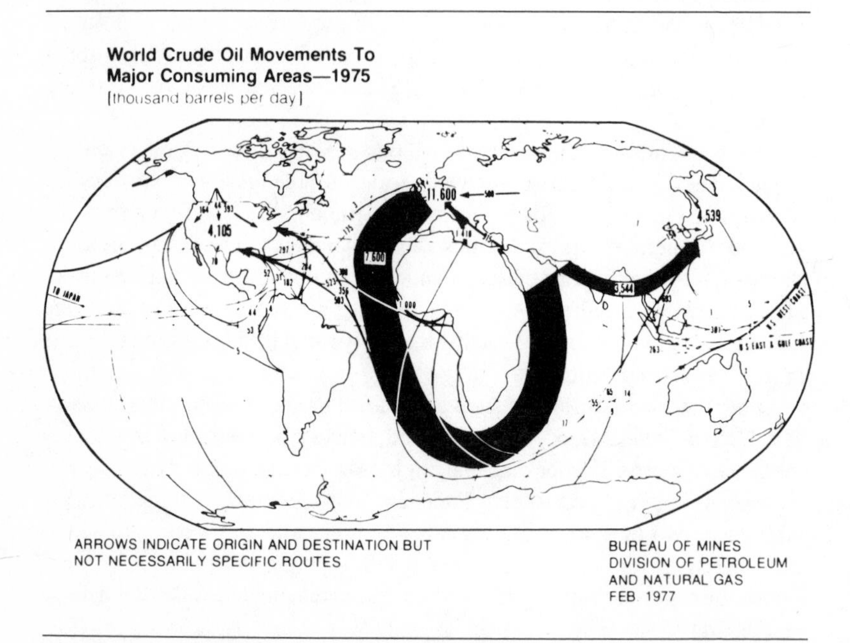


FIGURE 4.8: Relative American independence and European dependence
SOURCE: Bureau of Mines, U.S. Department of the Interior, February 1977.

Soon the United States was dealing "bilaterally" with the Arab oil producers like everyone else. Israeli diplomatist Abba Eban has even taunted the United States with becoming a "satellite" of Abu Dhabi and Dubai!

The disproportion between Middle Eastern and American oil production was so great as to suggest a new polarization in world affairs. Whereas the countries of the Middle East had an estimated 53 percent of known petroleum reserves, the United States had only about 5 percent. Represented visually by means of the contiguous-area cartogram, as in a series of Exxon Corporation advertisements, these statistics made a striking portrait of dependency. The coal situation, however, is quite different. The United States is believed to have the largest deposits in the world (Figure 9).[28] America's long-term energy "picture" thus would appear to be balanced, and its independence basically secured.

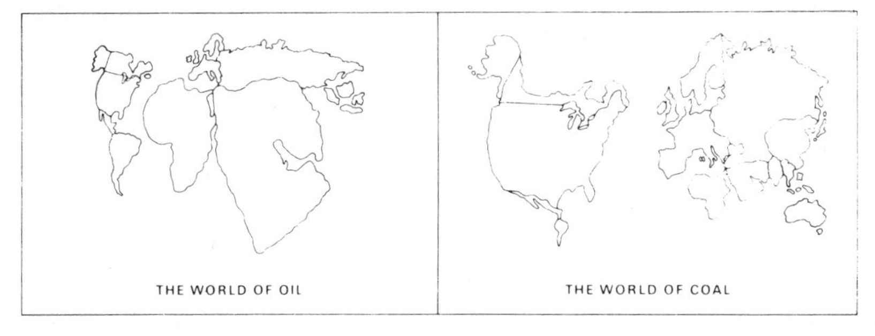


FIGURE 4.9: Oil insufficiency and coal sufficiency
SOURCE: Exxon Corporation and Mark S. Monmonier, Syracuse University

Nonetheless, the crisis of 1973 seemed to have a profound impact upon the hierarchical-locational structure of international relations and upon the position of the United States within it. Coupled with the policy of Détente and a nonideological, tilting conduct of alliances, it has introduced what may be termed a "new relativity" into international affairs. There are few absolute or fixed positions any more. What is center and what is periphery is no longer clearly apparent. Moreover, it is no longer even clear what dimension of international relations is most important. The result for many Americans, as mentioned at the outset, is a sense of geopolitical flux.

How can this uncertainty of Americans about their place in the world be understood? Several suggestive theories, or explanatory paradigms, of the fluidity of the international affairs have been offered. One is the "Diffusion of Power" thesis. According to it, the mastery of modern technology by more and more nations has reduced, and will finally eliminate, the advantages enjoyed by the industrial nations—including the United States. The gradients between economic centers and peripheries will thus be obliterated (Rostow, 1968). A second, more unusual, offering is the "Politectonic Zone" idea. Based on a loose analogy with the plate-tectonics theory in geology, this explanation regards political blocs as if they were freely floating on a liquid core. The accent is on nations' geological resources, geographical edges, and long-term chronological drifts. Most importantly for our present discussion, the position of any one bloc is relative to the positions of the others. None can logically be conceived as either central or peripheral (Cline, 1975).[29] A third, more cosmic, paradigm is the "Spaceship Earth" doctrine—the notion that, in the words of one of its leading proponents, "planet earth, on its journey through infinity, has acquired the intimacy, the fellowship, and the vulnerability of a spaceship" (Ward, 1966: v; Fuller, 1970). Patently, in such a world system, the physical, social, and psychic distances necessary to the maintenance of center-periphery hierarchies among peoples and places disappear. The United States, as the most distance-overcoming and altruistic of all nations, is usually expected, by Spaceship Earth theorists, to be the first to adopt this new selfless ethic.

Whether Americans can or will do so may depend upon their having a steady perspective and a firm sense of their own centrality. Is this objectively possible? The world position of the United States, though no longer supreme, is still a uniquely advantaged one. As Hassner (1976: 74–75) points out, "America alone is a decisive actor in every type of balance and issue." Lacking absolute power, it may be able to use its *relative* strengths to offset its *relative* weaknesses. To do this entails "the art of selectively separating or linking issues, regions and dimensions." Such a "differentiated" role, if played skillfully and with a singleness of vision, may enable the United States to hold the world center it has won in the short period of two centuries. If not, the American world position, and the international order still dependent upon it, may fragment and blur, like a cubist painting. From such a kaleidoscopic process, however, new, powerful focal centers may emerge—much as the United States itself emerged from a disintegrating world order.

## NOTES

1. The expression "central world power" appears to have gained currency following the change in U.S. policy or, in locational terms, reorientation toward mainland China—a more historic, traditional, and possibly self-assured "center of the world." See, for example, the argument of Naval War College professor Thomas H. Etzold, who, doubting that the United States is today "the center of international politics," maintained that it is "wrong to imagine that the United States is central, or even very important, in the larger concerns of the Communist

leaders of modern China" (Boston Sunday Globe, 1977). Compare the remark of George W. Ball, who, noting that Henry Kissinger visited Peking nine times, that two American presidents had been there, and that no leading Chinese official had as yet come to Washington, asked: "Are we, or are we not, vassals of the Middle Kingdom?" (Ball, 1977).

2. The confusion about the reason for the shift in the American world position—great-power competition or systemic instability—is partially indicated by a Harris Survey (1977) finding that, despite a sense of continuing rivalry with the Soviet Union, 60 percent of Americans think the United States "should work more closely with the Russians to keep smaller countries from going to war." For this reference, and for other assistance, I am indebted to John H. Maurer.

3. The pairs of correlative terms, "center"-"periphery" and "core"-"margin," often used interchangeably, differ somewhat. In general, center suggests a point, core an area. A center, unless a purely abstract, geometrical midpoint, tends to be dynamic; a core, though commonly a locus of strength, tends to be static.

4. Compare Robinson and Petchenik (1976: 74): "In mapping, one objective is to discover (by seeing) meaningful physical or intellectual shape organizations in the milieu *structures that are likely to remain hidden until they have been mapped.*"

5. The notion of a unique *central* sun did not last. As Kuhn (1957: 233) observes, "The center is the point that is equidistant from all points on the periphery, and that condition is satisfied by every point in an infinite universe or by none."

6. A more ornate astronomical analogy was drawn by Georges-Louis Leclerc, Comte de Buffon, who represented Britain as the "Sun," France as a "Comet." When the French comet, in its eccentric political movements, collided with the British sun, it rolled into orbit a new "Planet," "the American Empire" (Boyd, 1952: 443–445).

7. For examples of Philadelphia-based maps, see Lunny (1961: 38,43) and Ristow, (1977: 2). Examples of Washington-based maps are in Ristow (1972: 137, 146). In 1884 an International Meridian Conference, though held in Washington, fixed the prime meridian at Greenwich. For astronomical purposes, however, a Washington prime meridian (running through the Naval Observatory) remained legally in force until 1912.

8. For early cartographic representations of the Humboldt temperature-zonal system, see Robinson and Wallis (1967). See also "Gilpin's American Economic, Just, and Correct Map of the World," illustrating William Gilpin, *The Cosmopolitan Railway, Compacting and Fusing Together all the World's Continents* (1890). On the map a global linear "Isothermal Axis" is shown, roughly paralleled by the projected route of an intercontinental "Cosmopolitan Railway," connecting North America and Siberia at the Bering Straits. Reproduced in Karnes (1970: 150).

9. Among the earliest U.S.-centred world maps were David H. Burr's "The World, on Mercator's Projection," published by J. Haven in Boston in 1850, and "Colton's New Illustrated map of the World on Mercator's Projection," published by J. Colton in New York in 1851. I am indebted for this information to John A. Wolter, Chief of the Geography and Map Division, Library of Congress. Figure 2 is from Mitchell (1878: 4–5). All editions of Mitchell's *Atlas* from 1860 include this map, as I am informed by Richard W. Stephenson, Head of the Reference and Bibliography Section of the Library of Congress Geography and Map Division. A further, more intrinsically cartographic, reason for centering world maps on the United States at mid-century was the need to exhibit, in a unified way, the Wilkes Track, the route of the U.S. Exploring Expedition into the Pacific-Antarctic region (1838–1842). On a Greenwich-centered global chart, the Pacific Ocean is divided.

10. This position is shown figuratively in a Currier and Ives print, "The Stride of a Century," commemorating the 1876 Centennial. Behind the continent-straddling Uncle Sam figure, a

westward-moving railroad train carries the commerce of the Atlantic to the Pacific, where a ship awaits. Reproduced in Paterson et al. (1977).

11. The turn-of-the-century shift in American locational perspective is exemplified by the opening passage of Semple's (1903: 1) *American History and Its Geographic Conditions:* "The most important geographical fact in the past history of the United States has been their location on the Atlantic opposite Europe; and the most important geographical fact in lending a distinctive character to their future history will probably be their location on the Pacific opposite Asia." Semple's prophecy has some verification in current reality. The present American "global center," or the geometric focus of the smallest circle including all of the United States and its outlying areas, is in the vicinity of Kodiak Island off the southern coast of Alaska (U.S. Department of State, 1965: 1).

12. Other Europeans were more conscious of a direct American influence. Stead (1902: 19, 26) argued that within the "solar system" of English-speaking peoples the United States was becoming the "sun" and was able "to exert the pull" upon the United Kingdom. The increased Pacific orientation of Americans did not reduce the effect of proximity. As Stead noted: "Dublin is not half as far from New York as Manila is from San Francisco."

13. Compare Woodrow Wilson: "The United States has become the economic center of the world, the financial center. Our advice is constantly sought. Our economic engagements run everywhere, into every part of the globe." (Foley, 1923).

14. The principal author of the statement was the Italian emigré G. A. Borgese.

15. Figure 3 is from Spykman (1944 *The Geography of the Peace,* edited by Helen R. Nicholl, with an introduction by Frederick Sherwood Dunn, and maps by J. McA. Smiley.

16. Figure 4 is from Harrison (1944: 8–9), *Look at the World: The FORTUNE Atlas for World Strategy,* with text by the editors of *Fortune*. For a comprehensive examination, see Henrikson (1975).

17. See also Arnheim (1976). He regards a map "not as an assembly of shapes but as a configuration of forces," and stresses sensitivity to orientation. The images of countries on maps are transformed when rotated, as the United States would be if shown sideways or upside down. The "upright" position of the United States on polar maps minimizes (American) viewer disorientation.

18. For a map emphasizing polar routes, see Spykman (1944: 56).

19. Figure 5, adapted from the Mackinder (1943), is from Cohen (1963: 53).

20. For a conventional cartographic representation of the American alliance system, with lines radiating from Washington, D.C. see U.S. Department of State (1977a).

21. For graphic illustrations of the various U.S. defense perimeters, see Pearcy (1964).

22. For a recent historical discussion, see Gaddis (1977).

23. Figure 6 is from Weigert (1957: 274).

24. Figure 7 is from Morris (1973: 72–73). The maps were prepared by a *Milwaukee Journal* artist from originals by Don Clement of the *Los Angeles Times*. The virtues of Peters's map are extolled in Government of the Federal Republic of Germany (1977). Professional cartographers, discounting the novelty and technical merits of the Peters map (it has faulty standard parallels and a "squeezed" look), have been sharply critical of the excessive claims made for it. Nonetheless, it has served its provocative purpose.

25. "In a unifocal international system," as Liska (1967: 27) classifies the U.S.-centered world, "the relationship of any one state to the imperial state is operationally more significant for its role and status than is its position in a regional hierarchy and balance or its declaratory stance on matters of global concern."

26. This exaggeration and displacement of the Vietnam image are probably causally related, to a degree, to the repeated experience of seeing the map of Vietnam (without an

accompanying scale) on television screens. The image "filled" the screen, as did weather maps of the United States. Both countries occupied the same "space." The actual geographical relationship—distance and direction—of Vietnam and the United States was rarely, if ever, shown.

27. Figure 8 is from U.S. Department of State (1977b: 17). *The United States and World Energy: A Discussion Paper,* Publication 8904, General Foreign Policy Series 304 (November 1977), p. 17.

28. Percentage figures in the Exxon advertisement are from *Oil and Gas Journal* (December 25, 1972). Figure 9 is from Monmonier (1977: 19). On the official emblem of the Organization of Petroleum Exporting Countries (OPEC)—an oval shape showing all the members, from Ecuador to Indonesia, cartographically—the United States is not even on the map! (Allen, 1978: 23).

29. On a world map on the frontispiece, drawn to illustrate the Politectonic thesis, the "North America-Central America" zone appears to be floating upside down above Eurasia (the Heartland). For the notion of relativity in plate tectonics, see McKenzie (1977).

## REFERENCES

AGAR, H. (1940) The City of Man: A Declaration of World Democracy. New York.

ALLEN, L. (1978) "Not so wild dream: OPEC's amazing rise to world economic power." Harvard Magazine 80, 5: 22–28.

ARNHEIM, R. (1977) Letter to the author. June 9.

________(1976) "The perception of maps." American Cartographer 3, 1: 5–10.

BALL, G. (1977) "Against 'Cravenly Yielding' to Peking," New York Times (August 24).

________(1968) The Discipline of Power: Essentials of a Modern World Structure. Boston.

BARTLETT, J. (1970) The Record of American Diplomacy: Documents and Readings in the History of American Foreign Relations. New York.

BEALE, H.K. (1956) Theodore Roosevelt and the Rise of America to World Power. Baltimore: Johns Hopkins Press.

BOORSTIN, D.J. (1976) The Exploring Spirit: America and the World, Then and Now. New York: Random House.

________(1965) The Americans: The National Experience. New York: Random House.

Boston Sunday Globe (1977) August 21.

BOWMAN, I. (1928) The New World: Problems in Political Geography. Yonkers-on-Hudson, NY.

BOYD, J. [ed.] (1952) "Francis Hopkinson letter, January 4, 1784," in The Papers of Thomas Jefferson. Princeton, NJ: Princeton University Press.

CARTER, J. (1977) "Address at Notre Dame University." New York Times (May 23).

CLINE, R.S. (1975) World Power Assessment: A Calculus of Strategic Drift. Washington, DC: Georgetown Center for Strategic and International Studies.

COHEN, S.B. (1963) Politics in a World Divided. New York: Oxford University Press.

COOLIDGE, A.C. (1912) The United States as a World Power. New York: Macmillan.

CRESSEY, G.B. (1945) The Basis of Soviet Strength. New York: McGraw-Hill.

EICHELBERGER, C.M. (1977) Organizing for Peace: A Personal History of the Founding of the United Nations. New York: Harper & Row.

EISENSTADT, S.N. (1977) "Sociological theory and an analysis of the dynamics of civilizations and of revolutions." Daedalus 106, 4: 59–78.

FISCHER, E. (1948) The Passing of the European Age: A Study of the Transfer of Western Civilization and Its Renewal in Other Continents. Cambridge, MA: Harvard University Press.

FOLEY, H. [ed.] (1923) Woodrow Wilson's Case for the League of Nations. Port Washington, NY: Kennikat.
FULLER, R.B. (1970) Operating Manual for Spaceship Earth. New York: Simon & Schuster.
GADDIS, J.L. (1977) "Containment: a reassessment." Foreign Affairs 55, 4: 873–887.
GILBERT, F. (1961) To the Farewell Address: Ideas of Early American Foreign Policy. Princeton, NJ: Princeton University Press.
Government of the Federal Republic of Germany (1977) "Peters' projection—to each country its due on the world map." Bulletin 25, 17: 126–128.
Harris Survey (1977) June 9.
HARRISON, R.E. (1944) Look at the World: The FORTUNE Atlas for World Strategy. New York: Knopf.
HASSNER, P. (1976) "Europe and the contradictions in American policy," pp. 60–86 in R. Rosecrance (ed.) America as an Ordinary Country: U.S. Policy and the Future. Ithaca, NY: Cornell University Press.
HENRIKSON, A. K. (1975) "The map as an 'idea': the role of cartographic imagery during the Second World War." American Cartographer 2, 1: 19–53.
HOLBORN, H. (1963) The Political Collapse of Europe. New York: Knopf.
Inaugural Addresses of the Presidents of the United States from George Washington 1789 to John F. Kennedy 1961 (1961) Washington, DC: Government Printing Office.
JAMES, P.E. with E.W. JAMES (1972) All Possible Worlds: A History of Geographical Ideas. Indianapolis: Bobbs-Merrill.
KARNES, T.L. (1970) William Gilpin: Western Nationalist. Austin: University of Texas Press.
KENNAN, G. F. (1947) "The sources of Soviet conduct." Foreign Affairs 25, 4: 566–582.
KRAMNICK, I. [ed.] (1976) Common Sense, by Thomas Paine. Harmondsworth: Penguin.
KUHN, T. (1970) The Structure of Scientific Revolutions. Chicago: University of Chicago Press.
_________(1957) The Copernican Revolution: Planetary Astronomy in the Development of Western Thought. Cambridge, MA: Harvard University Press.
LaFEBER, W. [ed.] (1965) John Quincy Adams and American Continental Empire: Letters, Papers and Speeches. Chicago: Times Books.
LIPPMANN, W. (1952) Isolation and Alliances: An American Speaks to the British. Boston: Little, Brown.
_________(1947) The Cold War: A Study in U.S. Foreign Policy. New York: Harper & Row.
_________(1943) U.S. Foreign Policy: Shield of the Republic. Boston.
LISKA, G. (1967) Imperial America: The International Politics of Primacy: Baltimore: Johns Hopkins Press.
LUCE, H.R. (1941) The American Century. New York.
LUNNY, R.M. (1961) Maps of North America. Newark, NJ.
MACKINDER, H.J. (1943) "The round world and the winning of the peace." Foreign Affairs, 21, 4.
MAHAN, A.T. (1890) The Influence of Sea Power on History, 1660–1873. Boston.
MAY, E.R. (1961) Imperial Democracy: The Emergence of America as a Great Power. New York: Harcourt Brace Jovanovich.
McKENZIE, D.P. (1977) "Plate tectonics and its relationship to the evolution of ideas in the geological sciences." Daedalus 106, 3: 97–124.
MITCHELL, S.A. (1878) New General Atlas. Philadelphia.
MONMONIER, M.S. (1977) Maps, Distortion and Meaning. Resource Paper No. 75-4. Association of American Geographers.

MORRIS, J. A. (1973) "German's map of world improves on Mercator." Military Review: Professional Journal of the US Army 53, 11: 72–73.
PATERSON, T. G., J. G. CLIFFORD, and K. J. HAGAN (1977) American Foreign Policy: A History. Lexington, MA: D. C. Heath.
PEARCE, A. J. [ed.] (1962) Democratic Ideals and Reality, with Additional Papers, by Halford J. Mackinder. New York: Norton.
PEARCY, G. E. (1964) "Geopolitics and foreign relations." Department of State Bulletin 50, 1288: 318–330.
PRATT, J. H. (1942) "American prime meridians." Geographical Review 32, 2: 233–244.
RENNER, G. T. (1943) "Peace by the map." Collier's 113, 23.
RIPPY, J. F. (1958) Globe and Hemisphere: Latin America's Place in the Postwar Foreign Relations of the United States. Chicago.
RISTOW, W. W. (1977) Maps for an Emerging Nation: Commercial Cartography in Nineteenth-Century America. Washington, DC.: Library of Congress.
_________(1972) A la Carte: Selected Papers on Maps and Atlases. Washington, DC.: Library of Congress.
ROBINSON, A. H. and B. B. PETCHENIK (1976) The Nature of Maps: Essays Toward Understanding Maps and Mapping. Chicago: University of Chicago Press.
ROBINSON, A. H. and H. M. WALLIS (1967) "Humboldt's map of isothermal lines: a milestone in thematic cartography." Cartographic Journal 4, 2: 119–123.
ROSENMAN, S. I. [ed.] (1950) The Public Papers and Addresses of Franklin D. Roosevelt. New York.
ROSTOW, W. W. (1968) The Diffusion of Power: An Essay in Recent History. New York.
SEMPLE, E. C. (1903) American History and Its Geographic Conditions. Boston: Houghton Mifflin.
SMITH, H. N. (1950) Virgin Land: The American West as Symbol and Myth. Cambridge, MA: Harvard University Press.
SPYKMAN, N. J. (1944) The Geography of the Peace. New York: Harcourt Brace Jovanovich.
STANTON, W. (1975) The Great United States Exploring Expedition of 1838–1842. Berkeley: University of California Press.
STEAD, W. T. (1902) The Americanisation of the World, or the Trend of the Twentieth Century. London.
STEFANSSON, V. (1947) "The North American Arctic," pp. 215–265 in H. W. Weigert and V. Stefansson (eds.) Compass the World: A Symposium on Political Geography. New York.
SULZBERGER, C. L. (1973) An Age of Mediocrity: Memoirs and Diaries, 1963–1972. New York: Macmillan.
THROWER, J. W. (1972) Maps and Man: An Examination of Cartography in Relation to Culture and Civilization. Englewood Cliffs, NJ: Prentice-Hall.
Time (1977) "The master of the Mediterranean." 109, 21: 77–78.
_________(1972) "An interview with the President: the jury is out." 99, 1: 14–15.
U.S. Department of State (1977a) United States Collective Defense Arrangements. Publication 8909, General Foreign Policy Series 303. Washington, DC: Government Printing Office.
_________(1977b) The United States and World Energy: A Discussion Paper. Publication 8904, General Foreign Policy Series 304. Washington, DC: Government Printing Office.
_________(1965) United States and Outlying Areas. Geographic Bulletin No. 5. Washington, DC: Government Printing Office.
WARD, B. (1966) Spaceship Earth. New York: Columbia University Press.

WEBER, A. (1948) Farewell to European History, or the Conquest of Nihilism (translated by R. F. C. Hull) New Haven, CT: Yale University Press.
WEIGERT, H. W. (1957) Principles of Political Geography. Englewood Cliffs, NJ: Prentice-Hall.
WEINBERG, A. K. (1935) Manifest Destiny: A Study of Nationalist Expansionism in American History. Baltimore: Johns Hopkins Press.
WHITAKER, A. P. (1954) The Western Hemisphere Idea: Its Rise and Decline. Ithaca, NY: Cornell University Press.
WOHLSTETTER, A. (1968) "Illusions of distance." Foreign Affairs 46, 2: 242-255.

# 5

## REGIONALISM AND SOCIAL CHANGE IN ITALY

Francesco Compagna and Calogero Muscara

Although greatly increased efforts have been made in our time to lessen the weight of the differences and diversities that arise from the heritage of nature, and to reduce the inequalities due to the heritage of history, one can hardly say that the objective has as yet been achieved.

This persistent inequality, which is not necessarily linked to the diversification of the earth's surface, but often conditions it, results in widespread "geographic injustice," and to study this the concept of centre and periphery is a useful instrument. It is a flexible tool, whether one studies the different types of space that can be imagined (in that case, the concept is reduced to sectors of forces, sets of relations, economic space, and so on) or, in a more traditional but, alas, increasingly complex manner, historic organized space, which nowadays tends to coincide, even in political geography, with social space.

An accepted concept in geography, the idea of centrality, may be encountered as a basic condition (and sometimes a synonym) of many concepts and cognitive tools used in our research: from that of *crossroads,* which Jean Gottmann sees as the centre around which are woven the universal phenomena of circulation and movement characterizing the reality of our epoch, to that of the well-known theory of *central place*. In an economic view of space, centre and periphery may be associated with the *poles of development* that Perroux employs to introduce into economic analysis a closer conformity to the differentiations of reality. In political geography, it will suffice to cite Lucien Febvre's remark that "there is no state, no matter how small, that has not its vital nucleus, its geographic point of departure; there is no lasting political formation which does not owe its origin to a combination of forces, a kind of frame around which other territories will be able to grow, like flesh around a skeleton."

A wider use of the concept of centrality extended to its equivalent of periphery is to be found in the theme of the relations between city and country. This has led to the use of the conceptual tool, centre-periphery, in urban and social geography; one sees its trace in the stream of studies relating to Marxism—in those that apply the method of scientific analysis to political economy as well as in those that follow a theory of eschatological utopia (such as Coletti). The best-known ideological application of the Marxist concept of centre-periphery is that of Mao Tse-Tung, who has utilized the image of city and country as a synonym of the capitalist and underdeveloped worlds, and has attributed to China the international role of liberating the "countryside" from its condition of periphery of the world.

The following remarks seem called for: In a short essay it is not possible to reconstruct the long road followed by the idea of centre and periphery in the most recent theoretical endeavours of geography. In order to make a contribution to the validity of its use in setting forth and understanding very different geographic cases, one could try to examine specific examples of geographic situations introduced into the conceptual framework, with particular attention to the facts and phenomena that come within the purview of political geography. One could deduce therefrom indications of the possibility of using the concept of centre and periphery as an intermediary between systematic geography and regional geography, without losing sight of the complexity of reality. No less useful are indications regarding the flexibility of the concept as an intermediary with respect to other disciplines interested in territory and space—generally for applied research—and that make use of systematic methods and quantitative models.

Apropos to the usefulness of the concept of centre and periphery in political and even social geography, the Italian example is undoubtedly of special interest. For several decades there has been a process of change in the regional relations of centre and periphery within the nation, which evolved as a result of efforts to restructure the regions to fit the new social reality in Italy. A significant consequence of these attempts has been the emergence of a challenge to the tradition upheld since 1860 of the unit of the state.

## THE HISTORICAL CONFLICT

As is well known, the geography of Italy in its human as well as physical aspects is characterized by great local variety. North of Tuscany, these differences have deep roots in history and can be traced to the vitality of urban centres as well as to the inheritance of feudal states. In the rest of the peninsula and in the large islands of Sicily and Sardinia, despite the apparently greater uniformity of the Mediterranean mountain landscape and the fact that for centuries there had been only two important powers governing this territory, diversity is no less true. This variety owes as much to the difficulties of communication, which for centuries isolated towns and villages, as to the feudal social structure that survived in central and southern Italy until the beginning of the nineteenth century.

The extreme local variety of this human geography was the basis for two of the most important political propositions formulated in the course of the national *Risorgimento* for resolving the problem of the political fractioning of the country. One was the proposition of Mazzini, who saw the profound moral effort of founding the nation the most effective instrument for achieving political unity. Following on his proposition, Gioberti, who belonged to the Catholic milieu, and Cattaneo, a democrat, offered a federal state as a solution that would take into account, simultaneously, the aspiration of the ruling classes for unity and the deep-seated historic divisions existing within the people and between the states of the peninsula. To recall the political fractioning, one must merely remember that after the Congress of Vienna Italy was divided into a dozen states, bound by ties of different kinds to the major European powers.

The political line that won, leading in two years to the formation of the Kingdom of Italy, was that of Cavour. In his proposition, the processes of building an Italian state and of liberating the national territory from foreign powers had to coincide: Piedmont was given the historic task of defeating Austria to achieve national unity, though one should not neglect the role played by the volunteers of Garibaldi, who, landing in Sicily, reconquered the peninsula from south to north. Cavour was able to use the expedition of Garibaldi—a condottiere inspired by Mazzini—to give a popular aspect to the military campaign of the regular army of Piedmont, which lacked a political theme. Those who unified the young and frail Kingdom of Italy had, however, no illusions: Popular participation had been limited to a small class of young intellectuals and enlightened bourgeoisie. They were aware that there was no corresponding economic, social, or cultural unity in the anthropological sense. Consequently, Massino d'Azeglio was able to state: "Having made Italy, it was now necessary to make the Italians."

One of the first acts of the Kingdom of Italy was to resolve the debate between the regionalists and those who wanted a centralized state. This dilemma was resolved by instituting a highly centralized political organization, imitating that of the Kingdom of Piedmont, which itself had been inspired by the Napoleonic model.

It is far from certain that the consequences of the political unification led to attenuating the differences and inequalities inherited by the new state. On the contrary, in the economic domain a vast literature has given evidence of the price paid by the South after the political unification with the North, and how much the South has continued to pay since the North succeeded in developing its economy and joining the industrial economies of Europe.

But there is no doubt that the Italians of today are more alike than they used to be, not only because the world wars brought together populations which, until then, had no reason to meet, but also because the growth of bureaucracy has been an important factor in the migration of southerners toward the capital and northern Italy. The bureaucratic structure of the centralized state worked in the same direction.

It was not until after World War II and the institution of a new political regime that the theme of regionalism was again put forward. The left, with the exception of the Republican Party, hesitated to support the incorporation of regionalism in the new constitution. This happened because of the support of the Catholics on the one hand, and also because, during the war of liberation, regionalism had taken on the significance of a political renewal. It seemed to offer a way to resolve the contradictions through which political unification had been brought about. Moreover, in different parts of the country, and especially on the two large islands and the Alpine frontier, there were strong tendencies toward autonomy. The new Constitution of 1947 made provision for a regional organization of the state in addition to the administrative divisions already in effect in the provinces and communes.

A quarter of a century had to pass before the new regional organization was achieved throughout the country. At first, the only special regions created were those of Sicily and Sardinia in the South, and of the *Val d'Aosta, Trentino-Alto Adige*, and *Fruili Venezia-Giulia* in the North. These were areas where ethnic diversity had led to demands of autonomy for local communities of non-Italian language and for Italian-speaking communities holding customs and traditions different than those usually found on the Peninsula.

## REGIONS AND SOCIAL PATTERNS

Rapid industrial development, from 1950 to 1965, much greater than in earlier periods, made Italy a full-fledged member of the industrial countries of Europe. The effect of this on the land was to be seen in the expansion of industries in all of northern Italy and in a few places in the South. It also led to an integration into the consumption patterns prevalent within the framework of the modern cash economy and specialization, reducing the extent of the subsistence economy even in areas where, like the South, the industrial sector remained marginal. In this part of Italy there were attempts to establish several large basic industries and to undertake public works in the sectors of reforestation and soil conservation, road building, irrigation, and intensive agriculture of the Mediterranean type. However, initiatives to promote development proceeded with difficulty until the creation of a special technical-financial organization, the *Cassa per il Mezzogiorno*.

The most remarkable phenomenon in the South has been the mobility of the population. The migration of several million inhabitants to northern Italy and outside the country (four to five million to North Italy) has emptied the interior and caused the great urban centers to grow. This has resulted in a spasmodic but intense expansion of the construction industries.

All this could not fail to have an impact on the social transformation of the nation, increasing in particular the homogeneity of Italian society. Indeed, the "economic miracle" increased as never before the per capita income of Italians and established the material foundations for the creation of a new and vast middle class. Its way of life, characterized by mass consumption and urbanization, has become increasingly uniform.

But a strange and, in some ways, specifically Italian phenomenon has been that the process of diffusion of wealth and new customs has led the middle classes, including the civil servants, to adopt through the trade unions the anarchic-egalitarian ideologies of the student protests of 1968 in the eschatological-utopian version the Frankfurt school had grafted on traditional Marxism.

Before the student protests, Italian Marxism rested on an interpretation of the history of Italy elaborated toward the end of World War II by a young intellectual who studied under Benedetto Croce—Antonio Gramsci, secretary of the Communist Party. According to this interpretation, the backwardness and shortcomings of contemporary Italian society were due to the way in which the political unity of the country had been brought about. It had, in fact, been achieved "in spite of the Italians." It did not rest, according to Gramsci's theory, on the necessary national popular revolution, the only basis on which an Italian nation could be established and the precondition for a truly Italian state. Italian society had remained divided between the great mass of the poor and a small bourgeois minority that held all effective power despite the semblance of a formal and representative democracy. To redress the balance, the lower classes had to reverse the processes through which the country had evolved until then, with the subordinate and poorer classes, which the Italian state had neglected if not excluded from the modern life of the country, assuming the role of leaders.

The eschatological utopias of the post-Marxian left reinforced the view that sees Italian society as divided between the "real country" (*paese reale*), and the "official country" (*paese ufficiale*). The former designates the masses excluded from the decision-making process and the benefits of growth and development; the latter, the upper classes and their determination to keep their dominant role. The post-Marxian left extended the concept of subordinate class to the local communities and questioned the validity of the way in which the political unification of the country had been achieved. It believed that regionalism was the instrument for obtaining true democracy and direct participation, rather than the formalism of delegated and representative democracy.

This was one of the principal political theses in the programmes of renewal proposed by the left. In fact, behind this ideological mask is a hidden strategy for the conquest of power. The regionalism of the "historical left" in all its forms can be accounted for by the fact that the three regions of central Italy are controlled by a "red" majority. The more authority is granted to the regions, the more the parties of the "historical left" will be able to combat the power and authority of the state in Bologna, Florence, and Perugia (traditional centres of red domination), and today even in Genoa and Turin.

Against the background of the political and social changes that have taken place in recent years in Italy, regionalism has made great strides. In its support, the left-wing parties and the trade unions were able to make use of the mass media—newspapers and television—but also of publishers and the universities. The centralized structure of the political and territorial organization of Italy has entered a period of crisis. This has occurred not so much because of the weakening of the structure of the centralized state and the system of social values in what had been essentially a rural society, but also because this issue has been transformed by opposing the "real country" (the slums, local communities, and regions) and the "official country" (the centralized government, bureaucracy, and bourgeoisie) in a kind of class struggle.

The provisions of the Constitution of 1947 gave the regions authority in certain sectors (agriculture, tourism, local public transport, and vocational instruction) that were of relative importance. However, the attribution of these powers at a time of great social change enhanced their importance and has made of Italy an interesting case study in political geography with regard to the relationships between centre and periphery.

The question of whether the regional structure of the state corresponds to the polycentric geographic structure of Italian society could be elaborated through an examination of cultural differences and the spatial organization of the economy. The former are well illustrated by the linguistic diversity often described within the various parts of Italy. The latter takes on diverse aspects depending on the characteristics considered: opposition between mountains and lowlands, between rural and urban ways of life and traditions; unequal distribution of industry, income, and large-scale enterprise.

## THE DANGERS OF DISLOCATION

In Italy the idea of autonomy, if carried too far, whether applied to peripheral centres of local life or to regionalism, could have grave consequences for the future of the country. Admittedly, a responsible autonomy in the sector of decisions relating to economic enterprises, a proper defence of the accepted system of values and traditions, a love of one's land and dwelling place, can be excellent instruments (though not the only ones) for preventing the indiscriminate "march of cement" and for fixing the defence of an established spatial organization at points that already function as secondary growth poles. To this end, for example, one might choose the intricate and ancient urban web of northern Italy to prevent it from degenerating into a single, enormous, undifferentiated periphery of greater Milan. Of course, the question is one of the just measure; an excessive defence of local autonomy against all manner of coordination at a higher level could transform autonomy into a dangerous parochialism. In such a situation the local organization would then run the risk of being bypassed, then crushed, by forces larger than it, such as those of the large-scale capitalist economy operating in the heart of northern Italy.

The risk is present in the case of rural as well as urban regions. One can only view with favour a participation of the major regional authorities in the preparation of programmes for the *Mezzogiorno*. The great effort of mediation among the conflicting local interests that this requires provides a school in the practice of democracy and promotes an awareness of the web of regional interrelations in which each region is involved. However, to reduce all policy in favour of southern Italy to an ensemble of regional claims by the South could be harmful to the interests of the South, if the national programme is to coincide with the totality of the regional demands of the entire country. In one case, a policy favouring southern Italy might result simply in a mediation among the demands of the strongest regions instead of being a work of political synthesis, which goes beyond local claims. In the other case, the programme would represent only the interests of the strongest regions and would finish by widening the gulf between the two Italies.

A third and last consideration, which is no less alarming, concerns the nature of regional autonomy. All writers are in agreement in affirming that, as the case of Switzerland has frequently demonstrated, the more the administrative or political-administrative divisions of the state coincide with the local social life and its own traditions, specific set of values, and strong love of its *habitat,* the more the regional organization of the state will be workable. However, one must be aware of the risk of the economic differences between the regions weighing too heavily on regional policy and of the regions being pushed ideologically to invest too much in a direction contrary to the interests of the state.

The notion of nation, as Hartshorne has shown, includes a common objective. Playing the resentments of the Italian periphery against Rome in a direction contrary to the common objective of the nation means more than just a simple protest about the lack of efficiency of the central bureaucracy or about the marked difference in the ways, customs, and mentality of "province" and "capital," of "periphery" and "centre." One would even say that the fact of wanting to see in the regions an instrument for achieving a social revolution that would interrupt the political unification of the state borders on national disequilibrium. Little more than 100 years after national unification, there is great risk of returning to a reduced and fragmented country, such as existed before the Risorgimento; all the more so since, in the political debate raging over Italy in recent years, ideology often takes the place of a calm and rational examination of the actual political geography of the country.

If the conclusions to be drawn from our case study are clear with respect to applied geography, they are equally clear concerning general theory. The flexibility of the idea of centre and periphery, of which we spoke at the beginning of this chapter, has gradually assumed a somewhat ambiguous character. A deeper significance of the case we have studied is that one must go beyond a superficial interpretation. Regionalism adapted to social change does not replace the centre-periphery model, used in the sense of levelling differences and diffusing the points of decision and power. Such a proposal leads to a reversal of relations, in attributing all power to the periphery, and to removing power from the centre. On the theoretical plane, it suggests a reversal that does not eliminate the difference understood in the opposition between centre and periphery, but proposes to attribute the sense of centre to the periphery and vice versa. The inference from a geographic standpoint is that one must be very cautious with regard to formal structures that make use of the centre-periphery paradigm.

# REFERENCES

AMATO, G. [ed.] (1964) Il Governo dell'industria in Italia. Bologna: Il Mulino.

ARNALDI, G. (1974) "La storiografia come mezzo di liberazione dal passato," in F. J. Cavazza and S. R. Graubard (eds.) Il caso italiano. Milan: Garzanti.

BARBERIS, C. (1976) La società italiana classi e coste nello sviluppo economico. Milan: F. Angeli.

BELLAH, R. N. (1974) "Le cinque religioni dell'Italia moderna," in F. J. Cavazza and S. R. Graubard (eds.) Il caso italiano. Milan: Garzanti.

COLETTI, L. (forthcoming) Marxismo, Enciclopedia Italiana, Appendice IV.

COMPAGNA, F. (1967) La politica della città. Bari: Laterza.

CORI, B. (1977) "L'Italia: la geografia delle unità amministrative," in N. J. G. Pounds, Manuale di Geografia Politica, I. Milan: F. Angeli.

DEVOTO, G., and G. GIACOMELLI (1972) I dialetti delle regioni italiane. Firenze: Sansoni.

FEBVRE, L. (1922) La terre et l'évolution humaine. Paris: Renaissance du Livre.

FONDAZIONE AGNELLI (1972) Il sistema imprenditoriale italiano. Turin: Fondazione Agnelli.

FORTE, F. (1974) "L'impresa pubblica privata grande piccola," in F. J. Cavazza and S. R. Graubard (eds.) Il caso italiano. Milan: Garzanti.

GALBRAITH, J. K. (1967) The New Industrial State. Boston: Houghton Mifflin.

GAMBI, L. (1963) 'L'equivoco tra compartimenti statistici e regioni constituzionali. Faenza: F. Ili Lega.

GOTTMANN, J. (1947) "De la methode d'analyse en géographie humaine," Annales de Géographie.

GRAZIANI, A. [ed.] (1972) L'economia italiana, 1945–1970. Bologna: Il Mulino.

HARTSHORNE, R. (1940) "The concepts of 'raison d'être' and 'maturity' of states, illustrated from the mid-Danube area," Ann. Association of American Geographers.

KINDELBERGER, C. P. (1974) "Economia al bivio," in F. J. Cavazza and S. R. Graubard (eds.) Il caso italiano.

LE GOFF, J. (1974) "Il peso del passato nella coscienza collettiva degli Italiani," in F. J. Cavazza and S. R. Graubard (eds.) Il caso italiano.

LUTZ, V. (1962) Italy—a Study in Economic Development. New York: Oxford University Press.

MUSCARA, C. [ed.] (1978) Megalopoli Mediterranea. Milan: F. Angeli.

——— (1968) Una regione per il programma. Padua: Marsilio.

PERROUX, F. (1964) L'économie du XXéme siécle. Paris: P. U. F.

POUNDS, N. J. G. (1963) Political Geography. New York: McGraw-Hill.

PRACCHI, R. (1965) "I generi di vita della montagna italiana e le loro recenti modificazioni." Atti del XIX Congresso Geografico Italiano 2.

ROTELLI, E. (1967) L'avvento della regione in Italia. Milan: Giuffrè.

SCHUMPETER, J. A. (1954) Capitalism Socialism and Democracy. London: Allen, Unwin.

VALLEGA, A. (1976) Regione e territorio. Milan: Mursia.

# 6

# VARIATIONS IN CENTRE-PERIPHERY RELATIONS IN SOUTHEAST EUROPE

George W. Hoffman

The interaction between space and human society has always been a major concern of political geographers. Soja (1971) explained it as growing "out of the differential attributes of places, particularly those which act as major focal points of human activity." The importance of spatial interaction within the nation-state, especially between the "central nucleus and the outlying peripheral parts," is a concept much discussed by geographers, sociologists, and political scientists. Specifically, the meaning of the core or nuclear core concept has been analyzed by numerous geographers such as J. Gottmann, R. Hartshorne, S. Jones, D. Meinig, N. Pounds, E. Soja, D. Whittlesey, and others. Whittlesey (1944: 597) postulates state formation around a nuclear core "as an area in which or about which a state originates." In a later publication he (Whittlesey: 1954) points out that the core and the center need not be the same, in the sense that a nonnuclear core of original development may not be the focus of national life at a given point in time. Hartshorne (1950: 116) stresses the functional approach and points out that "a core area was neither sufficient nor essential to the evolution of a nation or state. . . .

AUTHOR'S NOTE: I wish to recognize the contributions to this study by my students, Joan Dee Held, Herbert V. Worff, and Leslie Wild, in a graduate seminar on political geography that used the theme of core-center/periphery relations. I also wish to acknowledge the use of material from my book, *Regional Development Strategy in Southeast Europe* (New York: Praeger, 1972). For the historical background of the following discussion, see Charles and Barbara Jelavich, *The Establishment of the Balkan National States 1804–1920. A History of East Central Europe*, vol. VIII (Seattle, University of Washington Press, 1977).

The common idea for a state may develop where no core areas exists." To him the purpose of the state is "integration" and this in the final analysis implies a state idea, a combination of centripetal forces with greater cultural force than the centrifugal forces acting in and on the area. Jones (1954) adds that effective integration is based on circulation throughout the state area in terms of communication and transportation. Pounds and Ball (1964), in their study of the European state system, also suggest that states generally develop through a process of accretion around a core. Finally, Meinig (1965: 213), in an analysis of spatial interaction and particularly the concept of the core-center, describes the function of the core as a center, pointing out that "a core, as a generic term, is taken to mean a centralized zone of concentration, displaying the greatest density of occupance, intensity of organization, strength and homogeneity of the particular features characteristic of the culture under study. It is the most vital center, the seat of power, the focus of circulation."

In spite of or perhaps because of the large amount of geographic literature, Burghardt (1969), among others, believes that there exists considerable confusion and a major ambiguity in this concept, especially in relation to its functional aspect. He argues that the term "core" can be viewed either as the *nuclear core* (Ratzel-Whittlesey concept—the cradle of the state), the original core (an area of concentration within an already larger political territory), or the *contemporary core* (the area, which at a specific historical period, is of greatest political, cultural, and economic significance).

Sociologists and political scientists also have examined the most appropriate use or meaning of the core or the center concept. Shils (1961: 117) stressed the center or the central zone as a locus of the realm of values and beliefs. "It is the center of the order of symbols . . . which govern society." Society, when viewed closely, appears to Shils to consist of a number of interdependent subsystems. Each of these subsystems comprises a network of organizations, and each of these organizations has an authority, an elite. As one moves from the center to the periphery over which authority is exercised, attachment to the central value system becomes attenuated. This remains as long as society is loosely coordinated, as long as authority lacks the means of intensive control, and as long as most of the economic life is carried on outside any market or nearly within local markets. On the contrary, contact with the central value system increases with the growth of the market and the administrative and technological strengthening of authority. This new contact creates a greater awareness and acceptance of the central value system, "while at the same time the new relationship increases the extent, if not the intensity, of active dissensus or rejection of the system. . . .

Thus, the periphery has ceased to be primary objects of authoritative decisions by others." (Shils, 1961: 129).[1]

Merritt (1974: 188) notes that "the core area of a region—whether defined in geographic, economic, or other terms—serves as the focal point for organizing the life of the rest of the region, known as the hinterland."

Rokkan (1973) uses a theoretical approach in the study of comparative political systems through the notion of territory. By studying the variations within the political system, he looks at the structure of space over which they exert influence. These are the "centers," which he defines as "gathering places where the major decisions are made . . . where the dominant actors in the system interact most frequently." Centers, according to Rokkan, control peripheries and their dependent populations by various transactions. The type of transactions prevalent and their relative emphasis in this center-periphery societal relationship has an important impact on the viability of the nation-state. It must also be made clear that both the center and the periphery, as well as spatial interaction between these two components, are in constant flux. Tarrow (1976, 1977) has done considerable work on core-periphery relations, especially in regard to France and Italy. His methodological approach to the formation of theory on the subject is through the criticism of three established models in an effort toward a synthetic model with policy relevance. These models contain three elements: moral hegemony, extractive coalition, and policy distribution; but these elements

> can vary in their combined effects according to the content of the ideology of hegemony of the elites, the composition of the ruling coalition, and the constraints and resources that obtain in the bureaucracy's policies toward the periphery. According to how these elements of the situation empirically combine, the periphery either becomes the supine victim of national economic and political colonization or marshals the resources to prevent itself from becoming little more than a collection of 'small towns in mass society' [Tarrow, 1976: 32].

Deutsch (1966: 39) perceives "centers" in terms of nodal areas of intense social communication of both facilities and customs "to the often much larger regions of hinterland which they may dominate in terms of transportation, strategies and economics."

In concluding this brief introductory discussion, it must be stressed that the geographic term "contemporary core" basically has the same meaning as the term "center" used by Rokkan and that "contemporary cores or centers" have a special relationship to their peripheries. I will only add, in reference to the area of Southeast Europe (the countries of Albania, Bulgaria, Hungary, Romania, and Yugoslavia), that in the relationship of the "contemporary core-center" to its periphery, the latter, at least in the initial stages of nation-building, usually encompasses the total territory under its control rather than peripherally located areas of the nation-state only, as defined in an earlier article by the author using examples largely from Western Europe (Hoffman, 1977a).

Political as well as economic activities, or simply certain physical geographic realities, often have contributed to the appearance of a number of nuclei, all with their own peripheral relationships, but Weigert (1957) argues that all ideas of geographical locations are of historical inheritance rather than functional realities. We should also add that while distance is today no longer an effective obstacle, in the past, as Gottmann (1975: 10) states, it separated "distant communities and preserved regional differentiation" resulting very often in different social and cultural characteristics. Soja (1971: 3) also considers distance a key factor in influencing "the differential attributes of places, particularly those which act as major focal points of human activity," though he considers distance not only as physical distance, but "both a socio-cultural 'distance' and what might be called functional 'distance.'"

One additional observation referring specifically to the spatial relations in Southeast Europe must be made. This involves the role of the traditional peripheral regions, sometimes referred to as the "mountain heartland." While normally considered as the "periphery," the mountainous heartland assumes considerable importance in specific periods of historical development as opposed to the sparsely populated, fertile plains and river valleys settled largely by foreigners, nonindigenous people often brought into the region for specific purposes, e.g., military defense and trading posts. This mountainous periphery, during certain historic periods in Southeast Europe, became the actual heartland and assumed the functions of a zone of security (see also the related discussions by Frank Trout), an area of retreat, and a refuge for the indigenous population fleeing from the invading armies. This population favored a stable social organization—stockbreeders and tribal associations—until recent times. The so-called periphery assumed the role of nation-building through key periods of history, e.g., the Vlachs, ancestors of today's Romanians; the Bulgars in the mountains surrounding the fertile Maritsa Basin; the Serbs in today's Yugoslavia; and the Albanians, to mention only a few. In Yugoslavia, because of the physical and economic barriers for the dominant ethnic groups seeking refuge from foreign invasions (Turks, Venetians, and Hungarians) in the mountainous heartland, they constitute a self-conscious region, causing in the period of nation-building in the nineteenth and twentieth centuries major obstacles to national integration, but at the same time preserving the state idea and thus ultimately contributing to the formation of a unified, though ethnically diverse state

(Hoffman, 1977b; Roglić, 1955, n.d.).

The center-periphery relations in the developing societies of Southeast Europe can only be understood if the various historical processes of nation-building in the region can be identified and the changing relationship between the center and the backward periphery can be analyzed. These changes can therefore best be explained by following the historical sequence of events: the center-periphery relations in the period of nation-building; the growth of contemporary centers and their impact on the periphery during the period of struggle for national liberation in the nineteenth century, which greatly influenced twentieth century problems; and finally the far-reaching structural changes with their center-periphery impact in the socialist countries of Southeast Europe after World War II. The changes in the postwar period in all of the nations of Southeast Europe, owing to the new social and political relationships that influenced industrialization, urbanization, and, to some extent, mass recreation, have again greatly affected center-periphery relations.

## CENTER-PERIPHERY RELATIONSHIPS IN THE NATION-BUILDING PERIOD

Historically, the process of nation-building in this area progressed very slowly and unevenly. The creation of the five existing nation-states in Southeast Europe was not an identical process and was only concluded in this century. None of these states can trace its state organization to a continuous, uninterrupted process.[2] The nuclei of several of these early states were often contested for the allegiance of diverse ethnic elements in shaping national consciousnesses. Some of these nuclei disappeared and later reappeared as a "contemporary core," often gaining power over ever-increasing peripheral regions around which states emerged. Geography and history have interacted in this area, bringing about a complex cultural landscape with great regional differences and with a "heterogeneous grouping of cultural variables and a political system that has been in constant flux as a result of external pressures and internal instability" (Fisher, 1966: 19). This is usually explained by its easy outer accessiblity, which made possible movements by many different people and ideas from neighboring power centers into important peripheral regions, and through its corridor-valleys into the heart of the region. Of even greater importance was the impact of the rugged and diverse relief of large segments of the area that produced greater inner fragmentation, encouraging particularism and isolationism and thus impeding important processes affecting the relationship between the growing centers of power and the backward peripheral regions.

The complex relief of the area is mainly responsible for the lack of political cohesion that made unification of the area extremely difficult and left a profound impact on the political and economic geography of the region. But the important role played by the mountains also was responsible for the fact that, in spite of the influx of competing foreign cultural influences, diffusion of these influences is superficial, with the result that Southeast Europe is an area not only in which different cultures are superimposed on one another, but in which people have been able to develop "distinct national personalities and highly articulated national cultures" (Valjevac, 1956; Burks, 1967: 82). The struggle against foreign domination was further complicated by ethnic and religious conflict, which resulted in the ardent nationalism for which the people of Southeast Europe are so well-known.

Into this highly complicated and diversified physical environment have come numerous people from many different physical and cultural environments. They have created an extraordinary and complex mixture of peoples of varying national consciousnesses that resulted in the area being well-known for its political instability, economic backwardness, and "distinct and relatively closed social system as well as sharply delimited communication system." (Lockwood, 1976: 4).

The core areas of the states of Southeast Europe generally have been classified historically following the well-known work by Pounds and Ball (1964). Hungary and Romania are classified as having expanded from a "nuclear area," though in the case of Hungary I would rather follow Burghardt's (1969: 350) view and refer to it as "original core" and not "nuclear core," in view of the absence of "evidence of accretion or growth from a 'kernel' over prolonged periods of time." Though Pounds and Ball note Szêkesfehérvár, the capital, as "nuclear core," a center of power including Transdanubia and, within a century, Croatia, Slovakia, and Transylvania was included in the Kingdom of Hungary and no further growth has been recorded since the twelfth century. The capital was moved to the hill of Buda overlooking the Danube in 1247, a site that developed into the modern core of the state.

While Romanian historians claim an uninterrupted process of growth from a "nuclear core," the province of Dacia of the Roman Empire, the first Romania state appeared in 1600 when the provinces of Wallachia, Moldavia, and Transylvania were united and ruled for a short time from Alba-Iulia. Wallachia, with its capital first at Târgoviste and later at Bucharest (1659), located in the loess-covered plain that later became the core of the modern state development, and Baia, Suçeava, and finally Iaşi (1565), 300 km. north, were the centers of power, but the scope of their power over the peripheral regions was never extensive. The modern Romanian state was created in 1859 when the union of Wallachia and Moldavia was achieved. Additional territory was obtained as a result of wars until it reached its present boundary at the end of World War I, though some territory was again lost at the end of World War II, especially the region of Bessarabia and Bukovina.

The South Slavic states of the Bulgars, Serbs, and Croats had numerous nuclei, generally weak in population, often rivaling each other, declining, disappearing, reemerging, and aiming to develop their separate national identities. The Slovenes of the North and of Macedonia succumbed after a short period of power to stronger neighbors. Over 500 years of Ottoman occupation of Bulgarian, Serbian, Macedonian, Albanian, Montenegrian, and Croatian lands, and consolidation of Austrian and Hungarian power in the northern parts of Southeast Europe, decisively influenced the relationship between the various centers of power and their peripheral regions.

The development of urban centers that could become centers of power, including focal points for national integration, was impossible in Turkish-occupied territory due to the makeup of the population consisting largely of foreign elements. The situation in the urban centers of the Austrian- and Hungarian-dominated lands was not radically different. While their administration was more efficient, the leadership was securely in Austrian and Hungarian hands. The objective of the newly built infrastructure was clearly oriented toward exploitation of the rich raw material and agrarian bases of the occupied region that served the commercial and industrial power centers of Vienna and Budapest. Local centers such as Zagreb, Karlovac, Split, and Novi Sad were either important headquarters of military commands or market towns serving a small hinterland, but the political and economic power located at these centers left the backward peripheral regions largely undisturbed. While small- and medium-scale industrial activities were being encouraged and were largely based on local raw materials, transactions of any importance between individual centers generally went via Vienna and Budapest.

Albania is a country without a distinct core area and no single center played more than a local market role. While its ethnic composition justified its independence, the establishment of the state was due to the power balance in that part of Europe during the first part of the twentieth century (see note 2).

## GROWTH OF CONTEMPORARY CORE-CENTERS AND THEIR IMPACTS ON PERIPHERIES

With the decline of the Ottoman Empire in the latter part of the nineteenth century, the struggle for national liberation became increasingly important.

While urban centers slowly gained in importance, their ties with the largely primitive peripheral, but generally densely settled, regions existing in isolation from each other scarcely improved.

The modernization of the economy and the broadening of the economic bases of the various regions of Southeast Europe were closely connected with the opening of the Ottoman Empire to foreign influences and the independence movements and subsequent establishment of the Southeast European states of Serbia, Bulgaria, and Romania. Economic contacts with the West began around 1860, and by the late 1880s a considerable amount of economic independence was reached. The need for capital for the economic expansion of these newly independent countries resulted in a slowly increasing foreign influence over all aspects of the economy and, at times, of the political life.

Many of the twentieth-century problems of the countries under discussion must be related to the period of nation-building in the preceding century and were mostly a legacy of the Ottoman Empire. This legacy of insuperable backwardness among the poverty-stricken peasantry created a serious problem of economic development and contributed to the great spatial differences and internal political problems, especially in today's Yugoslavia. The efforts made to establish a modern and often oversized bureaucracy, modern armies, a variety of mostly duplicating industries, and a transportation and public utility network, and the constant dream of its leaders of acquiring additional territories from its neighbors, creating unbridgeable rivalries among these countries, greatly influenced their economic and political development.

All through this period the main structural characteristics were changed little from the formative years and can be summarized as follows: a large labor force engaged in agriculture, with a high percentage of underemployed; extremely backward farming methods; a high rate of illiteracy among the masses, especially in those territories longest under Ottoman control; the continuation of an irrevocable trend toward further fragmentation of landholdings in agriculture, further aggravated by post-World War I land reforms aimed at providing farmland to the landless and underemployed, but that failed to accomplish this goal because of a constantly increasing population (Bicanić, 1955).

The drive toward industrialization actually had begun in the late nineteenth century. The formation of Yugoslavia from its nuclear core in Serbia with the inclusion of economically advanced Slovenia and the formation of Greater Romania, including Transylvania (both having important industries), acted as an incentive for further industrialization in the respective countries (Herz, 1947). But, in spite of all these efforts, the backward peripheral lands had little direct contact with the various regional centers. Even the newly established and generally isolated centers of economic activity—largely devoted to the exploitation of exportable raw materials—brought little change in the basic center-periphery relationship. The new ruling elite in the contemporary core had little contact with the peripheral areas of the state.

The nation-building period and the struggle for economic advancement went hand in hand. Extension of urban values from center to periphery and increased job opportunities were slow processes and closely tied to the spread of industrialization into the periphery. On the other hand, modernization and industrialization have had uneven impacts on countries' regions. In the short run this has contributed to regional consciousness. The basic problem is the central government's response in redressing the regional grievances, both in their political and socioeconomic aspects. These grievances encourage centrifugal forces within the state and can easily affect its cohesion. Greater mobility of population and spread of economic activity goes hand in hand with modernization and increases the contact between centers and peripheries. The establishment of a national capital for the newly organized state of Yugoslavia, and increased control over a larger territory by its capital in the case of Romania, resulted in increased power in terms of wealth and organization for these two capitals. In the case of Hungary, the reverse was true; a drastically reduced periphery region due to the loss of territory (Transylvania, Vojvodina, and Croatia) and the independence of the country resulted in an enhanced position of its Primate City, Budapest, which thus became a major center of population immigration, economic concentration, and political power. Widening disagreement between the rapidly growing centers and the periphery, to a considerable part, is attributable to the real or perceived economic exploitation of the latter. In addition, the initial economic development in the early period often resulted in increasing disparities between the more rapidly growing central nucleus of a country, its secondary growth center (regional capitals), and the outlying parts of the country.

A few urban concentrations in which industrialization was beginning showed some dynamic advances. They were the potential growth poles and centers, and it was hoped that they would lead to "spread" effects of modern economic development. Mihailović (1972), using the example of the East European countries and especially Yugoslavia, pointed out that polarization is to be expected at a low level of development simply because the volume of manufacturing is not great enough to spread over the whole territory. Industry is of necessity concentrated in regions where a superstructure, a skilled labor force, and a market exist. There were few such areas in the underdeveloped countries of Southeast Europe, and for this reason a certain amount of polarization in regional development was inevitable.

Basically, then, the emphasis in the development strategy pursued by all countries of Southeast Europe before 1939 was on changing the socioeconomic structure by industrialization, largely relying on foreign sources for the needed capital. Progress was extremely slow and at times, such as in the 1930s, did not exist or even regressed. A few larger industries located in potential growth centers, while on the increase, simply were unable to absorb the huge agrarian surplus population and lessen the plight of the backward periphery. The great spatial differences within individual countries, especially in Yugoslavia but also to a lesser degree in Romania, accelerated by the multiethnic character of their populations, remained unchanged and actually often were accentuated during the period of nation-building and continued all through the interwar years. The continued backwardness of large parts of the country resulted in increased peripheral unrest and considerable resentment about the rapid accumulation of wealth and power in some of the more important centers. The absence of any visible improvement in the backward regions resulted in increasing dissatisfaction and demands for equality in economic development. There was no doubt that the traditional economic and social structure was slowly changing, but what was really needed was a thorough reconstruction. The aftermath of World War II brought this thorough reconstruction, faster than anyone thought possible.

## CENTER-PERIPHERY RELATIONS IN THE SOCIALIST NATIONS OF SOUTHEAST EUROPE

Since World War II, few regions in Europe have experienced such far-reaching structural changes in their cultural, economic, and political life as the five socialist countries of Southeast Europe. Broadly speaking, the countries followed three closely related development aims throughout the postwar period: (1) a rapid economic development with industrialization as the main vector of change, thus hoping to absorb rural surplus labor in an increasing number of new industries and also to provide a rapidly growing number of secondary and tertiary employment opportunities; (2) the greatest possible use of domestic resources; and (3) an ultimate goal of "equality" for their backward peripheries. To accomplish these aims the whole development strategy of the postwar period emphasized rapid industrialization, which was to accomplish basic structural socioeconomic changes in the shortest possible time. According to the new leaders of the Southeast European countries, socialist development provided an opportunity, by the use of planned investments, to influence the location of new industries, thus contributing to a wider regional distribution and spread of economic activities, especially into the peripheral regions of individual countries (Hamilton, 1970).

Regional inequities in the five socialist countries of Southeast Europe, with their varied centrally planned economies (the Yugoslav system had undergone basic changes in the early 1950s that distinguished it from the other socialist countries), are numerous and complex. Development policies at various times have emphasized the development of underdeveloped regions by concentrating resource and development efforts under which all subnational units could be considered development regions.

The determination of long-term regional development policies in the socialist countries, which form the basis for middle- and short-term development plans, reflects societal goals of the country and ultimately aims to establish in spatial terms conditions that will reduce and eliminate regional inequalities in the standard of living for the population. The central authorities make their development decisions on the basis of the actual and potential performance of the economy and the need to maintain a fast rate of growth, and on the natural and demographic potentials and constraints, including the scarcity of resources. The interplay of these various forces has an impact on the structural formation of the economy, the strategy decided upon, and whether a particular region receives special attention in the development program of the country.

Another criterion in the economic planning of the socialist countries that influences regional development policies is the so-called "sectoral matrix." In the sectoral emphasis, the control authorities must reach a decision as to the location of new or the expansion of existing production as far as capital and infrastructure investments are concerned. That these decisions have a regional impact is obvious, but they clearly are not regionally disaggregated development policies. The authorities must decide between sectoral and regional priorities, as well as between decentralized and centralized emphases. Depending on the type of sectoral emphasis in a specific growth center the hinterland will be affected by the spread. In the socialist countries, slow growing and underdeveloped regions, regardless of their nationality, were singled out only in the later stages of their economic development. They receive assistance in the form of special investments within the overall development plan (Gruchman, 1975a, 1975b). Many of the regional imbalances, especially in Eastern Europe, have also been accelerated by past historical associations in various units of the new national states. The impact of these historical associations is still visible on the landscape and in the economic variables. While the communist governments have permitted ethnic groups (referring to cultural and symbolic characteristics of a particular society) to retain their cultures, including their distinctive languages, and have also seen to it that national minorities are always represented on regional and local decision-making bodies, ethnic differences (with the exception of Yugoslavia) are not regarded as significant criteria of social differentiation in societies in which relationships between classes are the main criteria.

Implementation of regional policy depends on the interaction between the decision makers and the economic system, with considerable differences in the type of economy existing: the Soviet type, centrally planned (command) economy, generally prevalent in Albania, Bulgaria, and, with some modifications, in Romania and Hungary; and the mixed (federal) type of economy prevalent in Yugoslavia, with decisions initiated and implemented by different levels of administrative organs.

If the experience of the Eastern European nations since 1945 provides a lesson it is that regional differences become no less sensitive simply because the economy of the nation as a whole may progress.

Two types of development strategies are discussed in connection with center-periphery relations: First, according to Mihailović (1972), development in the long run is connected with the creation of equal conditions. Thus disparities in regional development levels can be effectively attacked only when the national quest for maximum growth is tempered by a serious attempt to spread the growth among all regions of the nation. Instead of feeding ever-increasing amounts of resources into existing industrial centers, nations should strive for intensive development of these areas through higher productivity while establishing new growth poles in various underdeveloped regions.

It is important, too, according to Mihailović, that the right kinds of industry be established in the new growth pole areas. Ironically, some underdeveloped regions that contain significant natural resources have not fared well in development programs despite sizable investment. This is because investments in these areas have generally gone toward extraction or initial processing of raw materials. Such industries are capital intensive and not labor intensive, and, therefore, despite the large investments, few jobs are created. The development of the town of Bor in eastern Serbia (Yugoslavia), for example, operated the largest copper mine in Europe for 60 years with little effect on the general development of the local vicinity. Neither the iron and steel plants established at Niksić (Montenegro) as one of the first regional development projects in postwar Yugoslavia, nor the iron and steel plant established in Skopje (Macedonia), have created regional economic growth justifying the large capital investment and continuing budgetary subsidies required by these facilities. Certainly the original hope that these growth centers would in time diffuse "outward into the pole region and eventually beyond into the less-developed regions" has not always materialized (Nichols 1969: 193). The products of these industries are generally transported out of the backward areas to developed regions in unprocessed forms, thus with most value added and employment created at locations other than where they are most needed. Light manufacturing enterprises with high labor to output ratios generally have been more successful in spreading their economic-industrial activities.

Second, other scholars of regional development strategy would question the premise that all regions need to be developed; for example, Kuklinski (1970: 271) sees the main factor hampering successful regional development in the lack of real influence by the regional planning agencies in investment decisions; as long as sectoral ministries control investment decisions, regional considerations will always be subordinated to national priorities. He chides government officials for not recognizing that physical, cultural, economic, or social barriers in some regions might make them unsuited for industrial development.

Tarrow (1976: 36) raises the question whether "a diffuse system of territorial administration provides more chance for the periphery as a whole to capture public resources? Experience of federal systems such as Yugoslavia tells us that local power groups have more leverage at the state and local levels where administration is segmented than where it is partially integrated. But in centralized systems with diffuse control there is not even the moderate degree of autonomy found at the local and provincial levels of federal systems. Instead of being vertically segmented, policy-making is centered in the capital and dispersed among functional agencies.

Thus far attention has been drawn largely to the regional planning policies applied in the postwar period, especially those applied to regional inequalities directly affecting changes in the periphery. These policies emphasize in various degrees the regional allocation of resources, the coordination of national plans with regional objectives, and increasing consideration of the problems of lagging regions, as well as regions of overconcentration, with a greater involvement of the local population in the various decision-making processes.

Economic developments in the early stages often result in increasing regional disparities between the larger, often more traditional, development centers and the more backward peripheral regions. To encourage rapid industrial growth, the concentration of industrial production in few centers in the pre-socialist societies was due to the availability of a skilled resident labor force and market access. The same emphasis in economic development was given in the early postwar years and was the result of socialist central planning decisions with their emphasis on existing urban concentrations and their enlargement, on rebuilding from war damages, and on modernization of existing production. These developments especially stand out in the modernization process during the postwar years, which left its impact on center-periphery relations, industrialization, urbanization, and tourism.

## INDUSTRIALIZATION

Earlier it was emphasized that industrialization as the prime vector of change in the spatial distribution of economic activities was the mark of the postwar economic development in the region. The building of heavy industries (five new iron and steel works alone in different regions of Yugoslavia) received top priority, but their location at first hardly affected the periphery. Hand in hand with the forced pace in industrialization went a modernization and enlargement of key raw material exploitations. As new sites were developed, most of them located in the peripheral regions, they left a growing impact on the backward regions of Romania and Yugoslavia and, to a lesser extent, the other countries as well. This emphasis is today reflected in a greatly enlarged industrial base spread widely into the periphery, while in the earlier stages of industrialization, isolated factories and mines were developed near sources of raw materials. In the next step, development took place near consumer and service concentrations, usually at the traditional centers of a state. In a more advanced stage of economic development, industrialization became less dependent upon raw materials or existing markets and spread into the rural environment or small towns and the backward peripheral areas.

In some regions and countries, the establishment of new industrial activities in smaller towns brought about a fairly wide dispersion of economic and noneconomic activities: for example, in Slovenia, the northeastern parts of Bosnia and Hercegovina, Slavonia (between the Sava and Drava rivers), the southern part of Serbia, the Danubian Plateau, the Maritsa Basin, and the Black Sea Littoral of Bulgaria; Transylvania and western Wallachia, and central and northern Moldavia in Romania are also good examples. After some criticism about the early growth pattern in Bulgaria, resulting in concentrations in a few large cities—Sofia, Plovdiv, Dimitrovgrad, Pernik, Stara Zagora, Varna, and Pleven—the location of new industrial activities followed the principle of full dispersion, though polarization in a few centers had not been completely avoided. In some regions of Southeast Europe, industrial activities are concentrated in a few medium-sized towns, often the regional centers, that have a high potential as development poles. Such a policy of emphasizing development poles and growth centers comprising both economic and noneconomic activities is of special importance in Yugoslavia, where investment policies are highly decentralized and the need for increased industrialization in the backward peripheral areas is an urgent one.

In most of the Southeast European countries, regional equalization of the standard of living of the population now has top priority. To accomplish this goal the development of infrastructure has received much attention in the implementation of regional policies, since industrialization per se no longer plays an exclusive role and, according to G. Enyedi "an excessive regional dispersion of industries would contradict the principle of efficiency, so highly esteemed by the new economic mechanism" (Gruchman, 1975a: 68). While Enyedi's discussion refers specifically to the problem of Hungary, similar conclusions can also be drawn for Yugoslavia and, to a lesser extent, Bulgaria and Romania.

### URBANIZATION[3]

The rapid changes in economic activity and especially industrialization resulted in increased population concentration in the larger cities of the region and in a few secondary cities in every country, as well as a slow but steady increase in the total urban population. Bulgaria, Romania, and Hungary demonstrated the workings of Jefferson's (1939; Gottmann, 1975) law of the "primate city" with their important concentrations occurring in the capital cities of Budapest (20 percent), Sofia (11 percent of the country's total population), and, to a much lesser extent, Tiranë (8 percent) and Bucharest (7 percent). The federal character of Yugoslavia is clearly shown in the low population concentration of 4 percent in its capital, Belgrade. Even though restrictions were imposed in the early postwar years on moving to some of the capital cities, this slowdown brought only temporary relief. Yugoslavia, with its historical capitals of Belgrade for the state of Serbia and Zagreb for Croatia, has no primate city. Instead, both cities have experienced rapid growth, but do not play the overpowering role other key cities play in the region (Table 1). Several regional capitals and secondary cities are well distributed throughout the country and have assumed the role of growth poles and centers, offering an excellent opportunity for individual center-periphery contacts.

TABLE 6.1 Demographic and political data for countries in Socialist Southeast Europe

| | *Land area in 100 sq. km.* | *Population in millions* | *Density per sq. km.* | *Capital* | *Population of capital* | *Percentage of total population* | *Second largest city* | *Population* | *Year of official statehood and last boundary changes* |
|---|---|---|---|---|---|---|---|---|---|
| Albania | 28.7 | 2.4 (1974) | 76.6 | Tiranë | 216,000 (1974) | 9% | Durrës | 57,000 (1971) | 1920 |
| Bulgaria | 110.9 | 8.7 (1975) | 78.5 | Sofia | 962,500 (1974) | 11% | Plovdiv | 309,242 (1975) | 1920 1945 |
| Hungary | 93.0 | 10.5 (1974) | 112.9 | Budapest | 2,051,354 (1974) | 20% | Miskolc | 193,000 (1975) | 1920 |
| Romania | 237.5 | 21.0 (1974) | 88.4 | Bucharest | 1,565,872 (1974) | 7% | Cluj | 695,304 (1975) | 1920 1945 |
| Yugoslavia | 255.8 | 20.5 (1971) | 80.1 | Belgrade | 746,105 (1971) | 4% | Zagreb | 602,058 (1971) | 1920 1945 |

The basic pattern of postwar urbanization shows many similarities in each of the five countries due to the rural-urban migration and the growth of villages into towns. It is perhaps least developed in Albania, but the size of the country influences this development. The rapidly changing social and economic structure since World War II has considerable impact on rural areas with their traditional village structure and values, as well as their close family ties. The large-scale migration of younger people from the backward periphery (the mountainous heartland in Yugoslavia) to the rapidly growing, more viable regional centers or metropolises and the large-scale emigration from Yugoslavia (Hoffman, 1973) to Western Europe has had a lasting impact on traditions, values, and the agrarian structure of the countryside, but especially on the structure of the peripheral lands. Fuchs and Demko (1977: 62), in a recent study identifying the spatial population problems of the contemporary socialist economies of Northeast Europe (including Hungary), group them "into two major problem clusters which demonstrate the interface between spatial population and economic development processes: (1) the problem of regional disparities in rates of population and economic growth, levels of economic development, welfare, and manpower imbalances; (2) the problem of imbalanced urban networks including excessive urban population and industrial concentrations, disproportions in the rank-size distribution, and a lack of spatial integration." Clearly these two sets of problems are not mutually exclusive but are interrelated through migration, transport networks, investment allocation decisions, and other national-level processes.

The rapid urbanization brought about by industrialization resulted in a large-scale migration from the backward peripheral areas in all of the Southeast European countries. These migrations into the capitals and secondary growth centers created in many areas the necessary conditions for an accelerated center-periphery relationship. This was further encouraged by the rapid expansion of communications, thus bringing the outlying parts of a national territory within easy accessibility of the growth centers. Obviously this is an easier process in some of the smaller countries that have a developed infrastructure, such as Bulgaria and Hungary, than it is in Romania or even Yugoslavia (Mikhailović, 1969), with their long individual histories of various multiethnic backward regions (Hoffman and Hatchett, 1977). In addition, the great economic disparities between these regions under long Turkish and Austro-Hungarian control—roughly a north-south division—are even today the cause of considerable political and economic unrest.

Finally, it should also be pointed out that in recent years the center-periphery relationship was further enhanced by the increased industrial dispersion into the growing centers of the peripheral regions, especially for labor-intensive and footloose industries. Considerable urban growth was thus diverted to small and intermediate-sized cities. This deliberate policy of bringing industries to many peripheral regions has obviated to a large degree the need for long-distance travel among the working population. Initially, industrial growth enterprises absorbed surplus labor directly from agriculture and served as centers of economic growth that created additional employment opportunities. Thus growth in employment has spread over most regions. In Yugoslavia, growth in the officially designated underdeveloped areas generally exceeds the national rate. In most countries this resulted in a drastic reduction of the migrant labor and a corresponding increase in commuting. Its impact on a growing proportion of the rural labor force, which thus became engaged in nonagricultural activities, the so-called "peasant workers," was dramatic.[4]

## TOURISM

Economic well-being and technical advances contribute to an ever-growing mobility of the population and thus make increased vacation travel possible. Each of the countries of Southeast Europe has many beautiful spots to offer foreign as well as native tourists. (Albania, for political reasons, does not encourage visits of foreigners.) While some of these tourist spots are located in well-known and traveled areas, many are located in areas of difficult access with poor infrastructure until the most recent years. Modern hotels, private accommodations, and new access roads are rapidly spreading into the faraway peripheral regions, and the traditional sleepy market centers and villages have suddenly become booming tourist centers with important economic services for the numerous seasonal visitors.

The effect of tourism must be measured not only in its overall impact on a country's economy—this is of special importance to Yugoslavia with a 1976 tourist income of about $700 million from 5.6 million foreign visitors—but, perhaps more important, by its spread effect to the many smaller settlements. People from regions surrounding recreational areas and centers staff the various tourist accommodations, and the blossoming building industry (for infrastructure as well as accommodations) employs largely the local population. Many people from backward and often inaccessible villages find temporary employment during the summer, while keeping their permanent home and returning in the off season to improve their living accommodations using their summer earnings. Often young people and married men work in tourist resorts, returning to their family farms after the tourist season. As a result of this development, many peripheral regions are able to retain their populations and a new relationship is being established between the new tourist growth centers heavily depending on the seasonal tourist income and the periphery. Such a development has become of special importance to the poor and underdeveloped hinterland of the Adriatic littoral of Yugoslavia, but also to many interior locations in nearly every region of the country. It is of growing importance to the Rhodope mountain region, especially suitable for winter sports, and the rapidly spreading tourist region along the Black Sea littoral of Bulgaria as well as that of Romania (Varna in Bulgaria is an old, established tourist center, but numerous new Black Sea locations have become popular during the last fifteen years, both in Bulgaria and Romania). Backward areas in the Transylvanian Alps and the eastern slopes of the Carpathian mountains in Romania find themselves suddenly popular as tourist places, bringing much-desired income to a traditionally poor peripheral region.[5]

In conclusion it must be stressed that those peripheral regions offering recreational opportunity for more than one season obviously derive greater benefits. The new tourist centers rapidly assume an increasingly important role as growth centers, with the spread effect of their new, diversified, and mainly consumer-oriented industries having considerable impact on an ever-increasing area of the periphery. What is of importance and unique here, especially in the mountainous parts of Southeast Europe, is its impact on the traditional settlement pattern. Here the indigenous population remains in their traditional homesteads and regularly returns making improvements, whereas people usually leave permanently when working in the new industries of the rapidly growing towns and urban concentrations. Whether this relationship and concern for their traditional homesteads is a permanent development is difficult to say at the present time. Obviously, as long as tourism is the sole source of income, people will be interested in keeping close ties with their traditional homes; but if the new tourist centers assume the role of a broader growth center, their increasing growth and role will encompass the rural population.

What are the future trends in the center-periphery relationship, with economic criteria being put forward more frequently, especially in Yugoslavia and Hungary, as the sole factors that should influence the location of industries? Certainly the commitment to "evenness" in regional development may be waning. Such changes would likely be characterized by increasing interregional migration (in both Romania and Yugoslavia, intraregional migration has thus far predominated) from the underdeveloped peripheral regions to those with more favorable development potential. The result obviously would be greatly increased population concentration in certain areas (Figure 1), with the poorest regions generally losing many of their most economically active people and inhabited only by the old. To avoid such overconcentration, great emphasis is being exerted on channel-

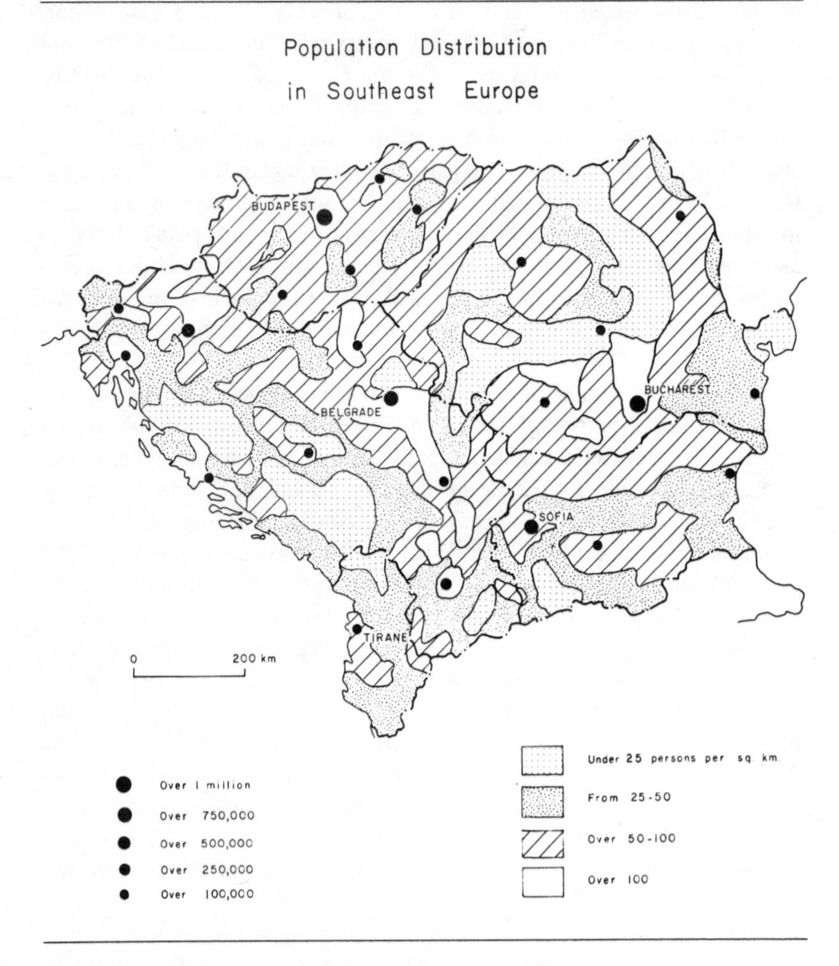


FIGURE 6.1: Population distribution in Southeast Europe

ing new industrial investments to an ever larger number of smaller and medium-sized growth centers, hoping for an ultimate spread effect on an increasingly larger peripheral region. On the other hand, the basic regional development problems for most of the countries of Southeast Europe continue to be avoidance of overconcentration, with expansion of industries and population into villages surrounding existing urban centers.

## NOTES

1. Alan K. Henrikson, in an aide-mémoires on the Paris meeting, raises an important question in connection with this statement, "Is not cultural leadership necessary—i.e., innovation as well as maintenance? Or does decision-making simply mean discrimination among existing values?"

2. The Albanian state was an exception. The Albanians, the probable descendants of the Illyrians, resisted most conquests and penetration and served Rome, Byzantium, and the Ottoman Empire. Their state was established by the European powers in 1913. Recent linguistic research seems to indicate that the Albanians may be a branch of the Thracians, who migrated from the province of Thrace, south of the Danube, to their present area; but available records do not indicate nation-building initiatives until the late nineteenth century.

3. There are numerous definitions of "urbanization" used in the countries of Southeast Europe. For details, see Kosinski (1974).

4. Fuchs and Demko (1977) noted the same for the northeastern part of Eastern Europe and the U.S.S.R. Recent studies show over 40 percent of the economically active males and over 30 percent of the females commuting to their place of work in Hungary, with one out of every four commuters away from home between 12 and 14 hours daily. Close to 200,000 economically active workers commuted to Budapest alone. Lockwood (1976) traces the history of peasant-workers and their extent and characteristics in postwar Yugoslavia.

5. For a discussion and map of the increased importance of tourism in Europe, including the countries of Southeast Europe, see Hoffman (1977).

## REFERENCES

BICANIC, R. (1955) "Five land reforms in Croatia (Yugoslavia) 1755–1953." (unpublished)

BURGHARDT, A. F. (1969) "The core concept in political geography: a definition of terms." Canadian Geographer 13 (Winter): 349–353.

BURKS, R. V. (1967) "Social forces and cultural change," in R. F. Byrnes (ed.) The United States and Eastern Europe. Englewood Cliffs, NJ: Prentice-Hall.

DEUTSCH, K. W. (1966) Nationalism and Social Communication: An Inquiry into the Foundation of Nationality. Cambridge, MA: MIT Press.

FISHER, J. C. (1966) Yugoslavia—A Multinational State. New York: ITT.

FUCHS, R. J. and G. J. DEMKO (1977) "Spatial population policies in the socialist countries of Eastern Europe." Social Science Quarterly 58 (June).

GOTTMANN, J. (1975) "The evolution of the concept of territory." Prepared for the Round Table of the I.P.S.A., Paris, January. (mimeo)
GRUCHMAN, B. (1975a) "Delimitation of development regions in centrally planned economies," pp. 1–13 in A.F. Burghardt (ed.) Development Regions in the Soviet Union, Eastern Europe and Canada. New York: Praeger.
________(1975b) "Key features of regional development and planning in Eastern Europe," pp. 259–268 in A.R. Kuklinski (ed.) Regional Development and Planning. Leyden: Sijthoff.
HAMILTON, F.E.I. (1970) "Aspects of spatial behavior in planned economies." Regional Science Association, Papers 25: 83–105.
HARTSHORNE, R. (1950) "The functional approach in political geography." Annals, Association of American Geographers 40.
HERZ, F. (1947) The Economic Problem of the Danubian States. London: Victor Gollancz.
HOFFMAN, G.W. (1977a) "Regional policies and regional consciousness in Europe's multinational societies." Geoforum 8 (September): 121–129.
________(1977b) "The evolution of the ethnographic map of Yugoslavia," pp. 437–499 in F.W. Carter (ed.) An Historical Geography of the Balkans. New York: Academic.
________(1977c) "Social and economic changes," pp. 125–132 in G.W. Hoffman (ed.) A Geography of Europe. Problems and Prospects. New York: John Wiley.
________(1973) "Currents in Yugoslavia: migration and social change." Problems of Communism 22 (November-December): 16–31.
________and R.L. HATCHETT (1977) "The impact of regional development policy on population distribution in Yugoslavia and Bulgaria," pp. 99–124 in H.L. Kostanick (ed.) Population and Migration Trends in Eastern Europe. Boulder, CO: Westview.
JEFFERSON, M. (1939) "The law of primate city." Geographical Review 29: 226–232.
JONES, S.B. (1954) "A unified field theory of political geography." Annals, Association of American Geographers 44: 111–123.
KOSINSKI, L.A. (1974) "Urbanization in East-Central Europe after World War II." East European Quarterly 8 (June): 129–153.
KUKLINSKI, R. (1970) "Regional development, regional policies and regional planning." Regional Studies.
LOCKWOOD, W.G. (1976a) "Religion and language as criteria of ethnic identity: an exploratory comparison (Austria and Yugoslavia)." Presented at the American Anthropological Association meeting, November. (mimeo)
________(1976b) "The peasant-workers in Yugoslavia," pp. 281–300 in B.L. Faber (eds.) The Social Structure of Eastern Europe. New York: Praeger.
MEINIG, D.H. (1965) "The Mormon culture region: strategies and patterns in the geography of the American West." Annals, Association of American Geographers 55 (June).
MERRITT, R.L. (1974) "Locational aspects of political integration," in K.R. Cox et al. (eds.) Locational Approaches to Power and Conflict. New York: John Wiley.
MIHAILOVIC, K. (1972) Regional Development. Experiences and Prospects in Eastern Europe. Paris: Mouton.
________(1969) "On the Yugoslav experience in backward areas," pp. 256–275 in E.A.G. Robinson (ed.) Backward Areas in Advanced Countries. New York: Macmillan.
NICHOLAS, V. (1969) "Growth poles: an evaluation of their propulsive effect." Environment and Planning 1.
POUNDS, N.J.G. and S.S. BALL (1964) "Core-areas and the development of the European state system." Annals, Association of American Geographers 54: 24–40.
ROGLIC, J. (1955) "Prilog Regionalnoj Podjeli Jugoslavije" (Contribution to the Understanding of the Regional Division of Yugoslavia). Geografskog Glasnika 16–17: 9–22.

———(n.d.) "Die gebirge als die wiege des geschichtlichen geschehens in Suedosteuropa. Colloquium Geographicum 12: 225–239.

ROKKAN, S. (1973) "Cities, states and nations: a dimensional model for the study of contrasts in development," pp. 73–97 in S.N. Eisenstadt and S. Rokkan (eds.) Building States and Nations. Models and Data Resources, Vol. 1. Beverly Hills, CA: Sage.

SHILS, E.A. (1961) "Centre and periphery," in The Logic of Personal Knowledge. Essays Presented to Michael Polanyi on His Seventieth Birthday—11th March 1961. London: Routledge & Kegan Paul.

SOJA, E.W. (1971) The Political Organization of Space. Commission of College Geography, Resource Paper No. 8. Washington, DC: Association of American Geographers.

TARROW, S. (1977) Between Center and Periphery: Grassroot Politicians in Italy and France. New Haven, CT: Yale University Press.

———(1976) From Center to Periphery. Alternative Models of National-Local Policy Impact and an Application to France and Italy. Occasional Papers No. 4. Ithaca, NY: Cornell University Western Societies Program.

VALJEVAC, F. (1956) "Die eigenart Suedosteuropas in geschichte und kultur." Suedosteuropa Jahrbuch 1.

WIEGERT, H.W. (1957) Principles of Political Geography. Englewood Cliffs, NJ: Prentice-Hall.

WHITTLESEY, D. (1954) "The regional concept and the regional method," pp. 19–68 in P.E. James and C.F. Jones (eds.) American Geography: Inventory and Prospects. Syracuse, NY: Syracuse University Press.

———(1944) The Earth and the State: A Study of Political Geography. Holt, Rinehart & Winston.

# 7

## CENTRE AND PERIPHERY: THE CASE OF ISLAND SYSTEMS

Lewis M. Alexander

As a geographical background against which to assess problems of "center and periphery," I have chosen island groups of the world ocean. One reason for my choice is the belief that political, economic, and social issues often show up more clearly, and are more readily measurable, in island groups, or "systems," than is the case on the mainland. A second point is that in the political-geographical world of today, islands are becoming an increasingly critical phenomenon—because of island-based independence movements, island-oriented economic development programs, and the new jurisdictional patterns being wrought by the advent of the 200-mile economic zone.

The growing importance of islands and island groups may be measured in several ways. About 30 percent of the countries that have become independent since World War II are island states, and the independence trend is still continuing. Indeed, most of the still-dependent territories of the world are islands. In economic terms a majority of the island countries are poor, many of them with few resources, limited areas, and small populations. Their strategic significance is evident from the air and naval bases at such sites as Guam, the Azores, Cyprus, and Diego Garcia in the mid-Indian Ocean, and from the location of island areas with respect to the coming assault on Antarctica. From a law of the sea perspective, islands take on an added meaning because of the projected 200-mile jurisdictional zones. One small islet, located 400 miles from its nearest neighbor, could close off 195,000 square nautical miles of ocean space within its own jurisdictional zone.

The term "island systems" should be clarified. A system may be defined, first, as "a group of interacting or interrelated elements forming a collective unit"; second, as "a network of structures forming communications or distribution." Both definitions have relevance for island environments.

Three forms of systems will be noted here. One form involves island groups that constitute a single state or territory. The islands of Oahu, Hawaii, and Maui are part of the Hawaiian system; several of the Line Islands, together with the Phoenix group, are outliers of the Gilbert Islands system.

A second type of island system is regional in scope and constitutes two or more political entities interrelated with one another through regional arrangements, such as those involving fishing, shipping, or economic integration. For example, CARICOM, in the Caribbean, is a regional economic arrangement, including most former and current British territories located there.

The third form of island system is a network of island areas, often far removed from one another and linked together by political, economic, or other ties. Such networks often include mainland territories as well. Historically, the British "lifeline" network to the Far East included the islands of Malta, Cyprus, Socatra, Ceylon, and Singapore; the current U.S. strategic network across the North Central Pacific involves Oahu, Guam, Luzon, and Taiwan. But networks may also be economic in nature, such as the Hawaii-Fiji-New Zealand commercial air link, or the air links across the Indian Ocean with stops at Mauritius or the Seychelles.

The pattern of island systems throughout the world is complex and varied. One primary division might be between coastal and "oceanic" islands, although just where the cutoff point should be between the two categories is a largely subjective issue. Some islands are little more than rocks; others are thousands of square miles in area. About 125 islands are greater than 1000 square miles in size. Some islands are part of large archipelago systems; others are associated with strung-out chains or small clusters; others are isolated units.

In addition to their size and distributional patterns, islands and island groups also vary widely in terms of population and economic development. Here one could compare Great Britain, Honshu, or Java with uninhabited cays of the Bahamas or atolls of the Maldives. Finally, island units differ considerably in terms of political status. Nauru, with 7000 inhabitants, is independent. The British Solomons, with nearly 30 times as many people, are not. It is, in fact, the political pattern of island systems, in terms of degree of self-rule and juridical relationships with one another, that makes the islands environment a particularly challenging one for political geographers.

To cut down the amount of island-related data to manageable proportions for this chapter, I have arbitrarily selected certain topics and eliminated others. I shall assess oceanic rather than coastal islands, and shall confine my study to island systems in the tropical and subtropical Pacific, the central and western Indian Ocean, and the Caribbean. I am not concerned with single island units, or with major archipelagos, such as Indonesia and the Philippines. It is in the smaller island systems that issues of center and periphery can be more readily analyzed. My procedure will be first to examine variations in the center/periphery syndrome of certain island groups of the study area, then to suggest some political consequences of these variations, assess potential regional organizations among the island groups, and finally identify certain "networks" that link island systems with one another.

## VARIATIONS IN THE INSULAR CENTER/PERIPHERY SYNDROME

### AN IDEAL SYSTEM

To start with, let us consider what might be seen as an "ideal" system in terms of a well-defined center with a contiguous periphery. A good example would be the Seychelles in the western Indian Ocean. The four major high-rising granitic islands are located within a 35-mile-wide area. Of the country's population of about 60,000, over three-quarters are located on the major island of Mahe, particularly in or near the capital city of Victoria. There are few ethnic differences among the population. The major airport is situated on Mahe and is served by such international carriers as British Airways, Air France, Air India, and South African Airways. The city of Victoria is also the major port of call for ships beyond the Seychelles, and encompasses the hospital, the teacher training college, and branches of international banks.

In addition to the four major islands, there are many small islands in the Seychelles, most of them uninhabited, stretching south toward Madagascar. Some of these were detached from the Seychelles system in 1965 and organized as part of the British Indian Ocean Territory; but after the Seychelles achieved independence in 1976, these coralline atolls were rejoined administratively to the new state.

Interisland communications within the Seychelles are maintained by air and water. Coconuts, spices, and more recently, tourism, form the basis of the economy. Although agriculture is carried on throughout the major islands, processing and export services are concentrated on Mahe. There are, in effect, no regional subcenters to rival Victoria. Activities radiate out from, but apparently seldom bypass, the well-defined national center.

One additional point to note about the Seychelles is that there are no other island systems nearby. The Comoros are over 800 miles southwest, Madagascar is 650 miles south, and the Chagos Archipelago is 1100 miles east. Clearly the Seychelles do not face an imminent threat of incorporation into some neighboring political unit.

There are a number of variations to this single-center model of island groups. One is where certain responsibilities are shared with the center by regional subcenters. Other patterns develop in which there are two or more centers, in which the center is relatively small, in which there are isolated peripheries, or in which island systems are jurisdictionally divided.

### REGIONAL SUBCENTERS

Examples of regional subcenters are found in Fiji, French Polynesia, and New Hebrides. In each case the capital city is very much the primary center, but there are some regional allocations of services and economic activities. In Fiji the town of Lambasa on Vannu Levu is the administrative capital of the northern division and has something of a cultural and commercial life of its own, exclusive of the influence of Suva, the country's capital and leading city.

For far-flung French Polynesia, each of the five "circumscriptions" has a district capital, although none of the smaller four in any way competes with Papeete, the territory's capital city. The international airport, the major harbor, and nearly two-thirds of French Polynesia's population are in Papeete. Distance is important here. Mangareva Island of the Gambier group is over 1000 miles east-southeast of Tahiti, while the northern Marquesas Islands are 900 miles northeast. In the case of New Hebrides there are four administrative districts, including three district centers in addition to Vila, the administrative capital and principal commercial center.

### TWO OR MORE CENTERS

The U.S.-administered Trust Territory of the Pacific has a series of centers. Saipan is the center for the Northern Marianas (as well as being the administrative headquarters for the entire TTPI). Kwajalein and Majuro are twin centers in the Marshall Islands, while the Carolines have four centers—Koro in the Palau group, Yap, Truk, and Ponape—spread out across 2000 miles of ocean.

The United States has divided the Trust Territory into six administrative districts. Of these, the Northern Marianas, separated as they are from the adjoining U.S. territory of Guam, are scheduled to become a U.S. Commonwealth in 1978. What will become of the Marshalls and the four districts of the Carolines remains to be seen. They might, for a time, become Associated States, but throughout the Pacific the pressures for independence are strong. In the Marshall Islands, Majuro is the district administrative capital, while Kwajalein is a U.S. Missile Base, and has acquired a large number of Marshallese defense workers and their families.

As will be noted later, there are multiple centers in the Lesser Antilles of the Caribbean, with no one focal point. And in at least two other cases, the administrative center is not the most viable economic site. One example would be the Gilbert Islands. Although Tarawa atoll is the administrative capital, the principal source of wealth for the colony is the phosphate production of Ocean Island, located across 300 miles of open ocean from Tarawa. There has been speculation that the phosphate reserves may soon be exhausted on Ocean Island, after which the 2000 inhabitants may be forced to relocate elsewhere.

The other case of off-center wealth is Papua-New Guinea, a polygot combination of island groups that became independent in 1975. The capital and leading city is Port Moresby on New Guinea, but a principal source of wealth is the copper deposits on Bougainville in the northern Solomons. Repeatedly one hears of incipient secessionist movements based on Bougainville.

#### RELATIVELY SMALL CENTERS

The Maldive Islands have a single center, but it is very small compared with the national periphery it must serve. The country consists of twelve clusters of coral atolls comprising over 2000 islands, spread over 550 miles from north to south. The 140,000 people of the Maldives live on about 200 of the islands, no one of which is larger than five square miles in area. The capital, Male, is located on Male Island, less than one mile across at its maximum extent. About one-tenth of the republic's population resides in the capital, and there are no other large towns to challenge it. But the Maldives are one of the world's least developed countries. How can health, education, and other services be adequately maintained over such distances, and from such a small city?

A somewhat analogous situation is provided by Tuvalu, the former Ellice Islands, which was separated from the British Crown Colony of Gilbert and Ellice Islands in 1975, and is tentatively scheduled for independence in 1979. The nine island-atolls are stretched out in a 400-mile chain. They have a total land area of 9½ square miles, and a population of about 6000. Some 800 people inhabit the 30 islets of Fanafuti atoll, where the country's administrative headquarters, airstrip, hospital, and jail are located. The island territory is poor, and with independence its government may be hard-pressed to furnish essential services to its peripheral areas.

## ISOLATED PERIPHERIES

In addition to variations in conditions of centrality among island systems, there are also differences in the peripheries as well. The Line Islands, for example, stretch for 1100 miles from north to south in the mid-Pacific; most of them are in dispute between the United States and the United Kingdom. Of the islands, only three are inhabited—Christmas, Washington, and Fanning. Administratively, these are part of the Gilbert Colony[1] whose capital is located 1800 miles west of the Line Islands. There are less than 1000 people on the three widely scattered islands. One alternative to continued association with the faraway Gilberts might be independence for the Christmas-Washington-Fanning group.

About 1500 miles southwest of Christmas Island are the eight Phoenix Islands, which are also the scene of political complexity. Although currently uninhabited,[2] they are claimed by both the United States and the United Kingdom. Most of the Phoenix group are administered as part of the Gilberts, but Canton and Enderbury Islands in the northern Phoenix are a U.S./U.K. condominium—an arrangement scheduled to last at least until 1989.

South of the Phoenix Islands are the three atoll islands of the Tokelau group, administered by New Zealand but claimed by the United States. Administration of the less than 2000 people of the Tokelaus is handled from New Zealand, there being no major centers of importance in the Tokelaus themselves.

There are other examples of "outlier" groups. In the case of the Cook Islands, which are a self-governing territory of New Zealand, there is a well-defined center, Rorotonga, in the southern sector, together with a number of smaller islands. But some 800 miles northwest are five islands of the Northern Cooks, containing about 2500 people and no central town. Four of these islands[3] are also claimed by the United States.

The Agelaga Islands of Mauritius in the western Indian Ocean are another example of outliers. The two small islands, located 600 miles north of Mauritius, have a population of about 500. The economy is dependent on coconut plantations, and the islands are actually closer to the Seychelles than to Mauritius.

## POLITICALLY PARTITIONED ISLAND SYSTEMS

Politically partitioned island groups are one variation of the center/periphery pattern. Two such groups are the Samoan Islands and the Solomons. In the case of the Samoans, the nine principal islands were divided at the end of the nineteenth century, with over 90 percent of the area and five-sixths of the population going to what is now the independent state of Western Samoa. Apia, Western Samoa's capital, is only 80 miles away from Pago Pago in American Samoa, but Western and American Samoa have relatively little interchange with one another.

The Solomons are divided between the northern islands (Bougainville and Buka), which are outliers of Papua-New Guinea, and the southern Solomons, with their capital at Honiara on Guadalcanal, which are a British protectorate. As noted earlier, Bougainville has copper and may in time represent a centrifugal force for Papua-New Guinea; however, should it leave the newly independent state, there is little to indicate that it would then join the southern Solomons, which themselves are headed for independence shortly.

An interesting political division occurs in the Comoros of the western Indian Ocean, where, in a plebiscite in advance of independence, the people of the island of Mayotte (with one-sixth of the Comoros' population) voted overwhelmingly to remain with France. The island group, lying at the northern end of the Mozambique Channel, thus remains politically divided.

A different type of political separation occurs among the U.S.-controlled western Pacific islands, where Guam in the Southern Marianas is an unincorporated territory, while the Northern Marianas, as well as the Marshall and Caroline Islands, are U.S.-administered trust territories. The island of Guam contains almost as many people as the entire Trust Territory and is clearly the dominant center of the area, although administratively separated from it. When the Northern Marianas achieved commonwealth status in 1978, their people, like those on Guam, became U.S. citizens, thereby reducing somewhat the barriers between the northern and southern parts of the island chain.

In the Caribbean, the Lesser Antilles, north of Grenada, are hopelessly partitioned. The British and French islands are interspersed, while at the northern end of the chain are the Dutch islands of Saba and St. Eustatius, as well as St. Martin, whose territory is divided between the Netherlands and France. There is no well-defined center in this chain of islands, nor do the independent states of Grenada, Barbados, or Trinidad and Tobago provide roles of centrality. The French islands of Martinique and Guadeloupe each have populations of a third of a million people, and the various British-owned territories combined have a third of a million people. The French islands are Overseas Departments, and the Dutch islands are part of the Netherlands Antilles, an integral part of the Netherlands. The British islands are Associated States, except for the colonies of Anguilla, Montserrat, and the Virgin Islands. The five Associated States[4] may before long become five independent countries, each with less than 100,000 people and with low levels of economic development.

## POLITICAL CONSEQUENCES OF CENTER/PERIPHERY PATTERNS IN ISLAND SYSTEMS

A first point to note with respect to island groups is the continuing trend toward independence. Most of the recently dependent territories on the continents have achieved this goal, and it is to be expected that during the next few years independence will be won by many of the island territories as well. Independence will often mean decreased administrative and other forms of support from the former mother country.

Along with independence may come some rearranging of the complex jurisdictional patterns of island units. Some groups, such as the Gilbert and Ellice Islands Colony and St. Kitts-Nevis-Anguilla, are already breaking up; in time French Polynesia and the Trust Territory of the Pacific may do likewise. And some rearrangements may eventually come in the Line Islands. But it is difficult at this time to anticipate any consolidations of island systems into larger political units. The history of independence over the past four and a half decades reveals very few examples of this trend.

### CONDITIONS OF CENTRALITY

Faced with the prospects, during the coming years, of a plethora of economically underdeveloped island systems with small populations and wide expanses of water between inhabited areas, the question of center sites becomes an acute one. Some island groups, such as Tuvalu and Yap, may have less than 10,000 people. Are there not some minimum requirements for national or territorial centers in areas such as these? And if the island units themselves cannot affort the necessary facilities, what alternatives exist?

One possible form of assistance to the sparsely populated island systems might be the development of regional centers: an airport capable of handling the largest jets, a medical complex, a deepwater port, and perhaps the headquarters of a regional university. These regional centers could be linked by sea and air with the national and territorial capitals, from which, in turn, links could be extended to outlying areas. With a little imagination, one could suggest a few such "centers"—Suva, Noumea, Guam, Tarawa, Papeete, Victoria, Port of Spain. But the effectiveness of these centers would depend on international regional planning, outside support, and a willingness on the part of island groups under different flags to cooperate in regional development. And such centers should not engage with one another in ruinous competition for trade or other advantages.

## CONDITIONS OF THE PERIPHERY

Because of the considerable differences among island systems with respect to their physical and cultural characteristics, it is almost impossible to generalize on the nature of peripheral interactions with the center. But certain categories of issues might be raised.

> (1) How strong is the island system's center in terms of its role as a centripetal force? Does it really affect the lives of the peripheral population? Does it provide the types of services they require?
>
> (2) What is the nature of the periphery? How far removed from the center are the other islands? Are there large "peripheral" islands or clusters of islets? Are there isolated islands far from the center?
>
> (3) How effective are the sea, air, and telecommunications links between the periphery and the center? A case might be made that the inhabitants of isolated peripheral areas, without adequate links to the center, might be better off being resettled on more accessible islands.
>
> (4) Perhaps most important, what are the perceptions of the peripheral inhabitants, so far as the island system's center is concerned? What have they been led to expect? Contrast, for example, conditions in the U.S. Trust Territory with those of the Gilbert Island or Tuvalu. In 1978 the United States is expected to spend about $1000 per inhabitant for administering the territory; per capita administrative costs for the Gilberts or Tuvalu are probably about 10 percent of that figure.

What differences might be found to exist in the center/periphery syndrome between island systems and countries on the mainland? There are extensive and sparsely inhabited countries of Africa, Asia, and South America. Barring the presence of mountains, it costs no more to fly 500 miles across the Pacific than across the Sahara; in both cases the area traveled may be virtually empty of inhabitants. In the place of shipping lanes, there are roads, caravan routes, pipelines, and perhaps, railroads. But the person in Timbuktu is as far removed from Bamako as is the inhabitant of the Bismarck Archipelago from Port Moresby, or the people of the Austral Islands from Papeete.

One could discourse indefinitely on the physical differences in the center periphery syndrome between land and sea areas; what is more essential, of course, are the differences among the peoples of the two environments in their perceptions of their relationships with the center. In fact, do spatial perceptions in general differ among oceanic island peoples as opposed to those surrounded completely, or almost completely, by land? Here is a question I have neither the space nor the wisdom to pursue, but perhaps as "marine" or "island" geography develops, some progress may be made toward addressing the issue.

## REGIONAL ORGANIZATIONS AMONG ISLAND GROUPS

The development of multistate regional organizations has taken place along two lines. One is essentially geographical in nature, involving a variety of functions within a given area. Such, for example, is the European Economic Community—a decidedly unique feature on the world scene in terms of both its success and the scope of its competencies. The other category is single-function units, such as those involved with river basin development, fisheries, or marine science research. There are, for example, twenty-seven regional fishery bodies in existence, seven units associated with regional maritime development and five arrangements for environmental control in regional marine areas. Institutions of these types affect island systems and serve to bring them together more closely as regional units; but the institutions are not directed primarily toward interisland cooperation, and often do not constitute strong centripetal forces for interisland unity.

The difficulties experienced by "geographical" arrangements on land (such as the East African Community or the Andean Group) are transferable to island systems. The same petty jealousies are there, the same differences in perception of the costs and benefits of regional action. Nevertheless, two regional island systems have emerged which I would like to note briefly—CARICOM (Caribbean Community and Common Market) and SPEC (South Pacific Bureau for Economic Cooperation).

Established in 1973, CARICOM encompasses all of the British territories of the West Indies except the British Virgin Islands, Belize, and the independent (former British) states of Barbados, Grenada, Guyana, and Trinidad and Tobago. The CARICOM Treaty provides for the coordination of the foreign policies of member states, the economic integration of member states through the establishment of a Common Market, and functional cooperation through the creation of such institutions as a Development Bank, an Investment Corporation, and a regional shipping line. Although CARICOM is only beginning to get under way, it represents a hopeful step in the direction of multipurpose regional cooperation among both dependent and independent island communities.

Recently formed and with headquarters in Suva, SPEC membership includes Fiji, Western Samoa, Tonga, the British Solomons, Niue, the Cook Islands, the Gilbert Islands, and Tuvalu, with Australia and New Zealand as supporting members. It is an offshoot of the South Pacific Commission, which includes nonregional members—France, the United States, and the United Kingdom. At this time SPEC is primarily concerned with fishing and other aspects of maritime development, but it could develop into an important multipurpose organization.

It is difficult to separate the center/periphery syndrome from issues of multistate regionalism. This is particularly true with respect to island systems, since these are so often weak in terms of economic and cultural development. But even with regard to maritime activities, regional action at this time is difficult because of the advent of the 200-mile economic zone, which gives each island community exclusive control over living and nonliving marine resources, scientific research, and pollution control to a maximum distance of 200 nautical miles from shore. Faced with these new acquisitions of ocean space, some island groups have been reluctant to enter into regional fisheries or other arrangements that they perhaps see as jeopardizing their recently won rights.

## INSULAR NETWORKS

In the case of island networks, one should first note the pattern of imperial systems that took shape following World War II, when the island dependencies of Britain, France, Spain, Portugal, the United States, and other powers were spread across the world ocean, forming a series of political networks. In the more than three decades since the war, Britain has divested itself of a considerable number of island territories and now seems determined to accelerate this independence process as much as possible. The other powers have, at a slower pace, granted independence to their island territories. Some of the former dependencies, such as Trinidad and Tobago, the Maldives, and Papua-New Guinea, have retained close network ties with their former sovereigns; others, such as Indonesia, have not. And there are still a considerable number of island systems that remain a part of the empire networks, and thus a part of economic and, frequently, military networks based on the mother country.

The case of strategic island networks is one that has changed considerably over time. Not only do such networks include air and naval bases, but also missile tracking stations and communications centers. The U.S. trans-Pacific network through Hawaii, Kwajalein, and Guam has been augmented by facilities on Midway and Johnston Islands, as well as on several of the Line Islands. Because of overcrowding on Guam, some experts believe that ultimately nearby Rota and Saipan Islands may be pressed into military use, and there may well be need for staging fields or refueling or communications facilities at Ulithi Atoll in the Yap Group in Micronesia, as well as in the Palau Group. But former military facilities in American Samoa and on Christmas Island have reportedly been closed down.

Britain's strategic island network used to extend east of Suez, and the closing down of the British base at Gan was eventually followed by a U.S. decision to open a base on Diego Garcia, 400 miles south-southwest. France maintains a naval presence in Martinique, Diego Suarez (Madagascar), Reunion Island in the western Indian Ocean, and French Polynesia. The press frequently contains reports of prospective Soviet air and naval bases on islands in the Atlantic, Pacific, and Indian Oceans. The Soviets assisted Iraq in the constuction of the harbor facility at Umm Qasr, and India with the naval base at Vizekhapatnam; they helped with the harbor clearing at Chittagong, Bangladesh, and are reportedly interested in the use of port facilities at Port Louis, Mauritius. Although recently ousted from Berbera, Somalia, they apparently have access to port facilities at Aden, Southern Yemen. But there is little solid unclassified data on the development of strategic island networks, and the student of this subject must piece information together from a variety of sources.

A third category of island networks is economic in nature. Among this type are those concerned with transportation, tourism, and fishing. There is a worldwide commercial air network, involving such islands as the Canaries, Azores, Cape Verdes, Mauritius, Seychelles, Fiji, and Tahiti. Large international airports, such as Nadi, Fiji, become centers for regional air traffic, linked to a global commercial air net. Similarly, in shipping, certain groups have become part of the supertanker or containership nets. Among the supertanker ports are those of the Bahamas, Heligoland, the American Virgin Islands, Okinawa, and Newfoundland. Along with the vessels come shipping agents, distributors, and at times, industrial plants. And here, as in the case of air transport, regional island centers are developing as part of far-flung networks.

In the field of tourism, we now have the resort hotel chains. Among these are Sheraton, Hilton, Western International, Continental, and Inter-Continental, not to mention Holiday Inn, the Princess Hotels, and Caribbean International Resorts. Many of these chains have interconnecting arrangements, so that a prospective tourist in Europe or North America can use a worldwide network of comfortable and well-appointed island resort hotels linked with one another by air/sea connectors.

As far as fishing is concerned, the principal fleets using island bases are those in whaling and tuna. As restrictions on foreign fishing increase within coastal states' 200-mile zones, the emphasis on species of the open ocean will increase, and other fisheries resources, such as billfish and lantern fish, may become subject to greater harvesting. More important, perhaps, is the harvesting of the Antarctic krill, whose estimated biological potential may be as much as half that of presently used species of the world ocean. To support the fleets for these distant-water efforts, networks of supply bases may increasingly be established on mid-ocean islands.

Organizing islands for political, economic, military, or other purposes is a highly complex process. Traditionally, political geographers have tended to focus on political forms and processes associated with continental land masses, while marine affairs specialists concentrate on water areas. But islands and island systems are a part of the marine environment and are intimately tied in with any management programs drawn up for that environment. Much work is needed by geographers and other social scientists in identifying and assessing the changing nature of island interactions as they affect the international political scene.

## NOTES

1. Christmas Island is currently in dispute between the United States and the United Kingdom.
2. Some years ago, about 1000 people from the Gilbert Islands were settled in the Phoenix group, but efforts to become economically independent were unsuccessful and they were subsequently resettled in the British Solomons.
3. Danger Atoll, Manahiki, Rokahanga, and Penthyn.
4. Antigua, St. Kitts-Nevis, Dominica, St. Lucia, and St. Vincent.

# 8

## THE CITY CENTRE AS CONFLICTUAL SPACE IN THE BILINGUAL CITY: THE CASE OF MONTREAL

Jean A. Laponce

In the most ordinary forms of the language as in the most technical, in social science analysis as in mythological and religious symbolism, the contrast between centre and periphery appears again and again with remarkable regularity. Sometimes the contrast is used to describe geographical properties, sometimes to describe social and cultural characteristics of a nonspatial character. For Christaller (1966), the distinction refers to physical space; for Shils (1975), it refers to institutional relationships; for Rokkan (1970; Eisenstadt and Rokkan, 1973–1974), it refers to both.[1] The ease with which we slide from physical to metaphorical space is a source of great confusion—a confusion so widespread as to call for an explanation. That explanation—which resides in the mutual attractiveness of the notions of height and centrality—will remind us once more that, whether studied from the vantage point of geography or from that of political science, human behaviour is best understood if we do not divorce behaviour from perception.

Taking linguistic conflicts in the city of Montreal as an example, I shall argue the usefulness of analyzing ethnic tensions as a function of an individual's mental maps. In addition to considering the relevant census information on linguistic divisions in Canada, Quebec, and Montreal, this chapter will consider the relationship between the archetype of verticality and that of centrality, it will point to the need for political science to emancipate itself from too exclusive a dependence on "objective" administrative boundaries in order that it may consider as well the subjective perceptual containers that give meaning to the behaviour of political actors, and it will relate linguistic to social and territorial hierarchies. Linking these various and at times seemingly disparate considerations, there stands a simple question and a tentative answer. The question is whether the modern multilingual city can become and remain as easily multilingual in its centre as it is at the peripheries. The tentative answer is that linguistic diversity can be more easily maintained at the peripheries than at the centre, with the result that a city that used to be tolerant of multilingual diversity in its phase of mobilization becomes intolerant in its phase of consolidation; tolerant when enough of its inhabitants were still oriented to the hinterland from which they came, intolerant when enough of these same inhabitants look inward, look increasingly toward the same nondivisible city centre. For lack of empirical data, I must often speculate—and these speculations should be read for what they are: a call for data.

## CENTRALITY AND DOMINANCE

The notions of a centre within a circle or a centre at the intersection of a cross—whether it be in their abstract geometric purity or in the vague notions of crossroads confronting us and of environments wrapped around the self—are among the three or four most powerful spatial archetypes. They are so powerful in generating and constraining our thoughts that it appears doubtful that we could dispense with them.

A centre that separates self from others, the in from the out, the sacred from the profane, the powerful from the weak; a centre that integrates and often also liberates; a centre that rules, governs, and dominates—such notions appear in all cultures. Gottmann (1974) contrasts two models—that of Plato, focused on a centre that dominates peripheries that are used to isolate; and that of Alexander, which values equally any location in the system and uses the peripheries as openings on the outside.[2] Of the two, the first is that which continues to dominate our thoughts even after the cosmos ceased to turn around the earth. Whether the privileged attention given to the centre is due to the fact that man's vision is clearer in the middle than at the edges of the visual field, whether it is due to our being constantly aware that the environment is always all around us, whether it is due to the self-centredness resulting from our having better control over what is proximate than over what is faraway—whatever the reasons, we always keep searching for centres, geographical as well as metaphoric, and we keep linking the two. This need for a centre appears in the complex ideologies analyzed by Paul Mus,

Mircea Eliade, and Claude Lévi-Strauss as in the most trivial of intellectual mappings.[3] Consider, for example, an experiment by Peckjak (1972), who asked manual workers in two markedly contrasted cultures—those of Yugoslavia and Zambia—to locate a series of very positive and very negative concepts (friend-enemy; happiness-sadness, and so forth) in a variety of suggested spatial orders that included centre-periphery, vertical up-down, diagonal up-down, left-right, and so on. He found the centre-periphery contrast to be the one most universally applied in both cultures and the one that contrasted and predicted best the relative locations of the negative and positive concepts classified by his respondents.

Other simple experiments can be used to verify the greater visibility and the perceived uniqueness of a central location. Consider the following: I asked 53 students (interviewed as a group in their regular classroom, but answering individually and in writing) to make a choice for one of the nine dots displayed as follows:

| • | • | • |
|---|---|---|
| • | • | • |
| • | • | • |

The subjects were told that their answers would be matched at random with those of other subjects taken from another class, and if their answer matched the other's answer, they would receive a modest prize ($2.50).

Measured in percentages, the distribution of choices was as follows (N = 53):

| 8 | 12 | 17 |
|---|---|---|
| 6 | 37 | 8 |
| 4 | 6 | 10 |

As expected, the centre location was the one most often selected, but I was surprised that it had not been chosen more often.[4] Three weeks later, before announcing the results of the original test, I asked the same respondents—who did not know with what group their responses had been matched—to recollect their choices and to indicate which of the dots they now thought was most likely to be a winner.

Measured in percentages, the recollection of one's original choice gave the following distribution:

| | | |
|---|---|---|
| 11 | 6 | 9 |
| 6 | 40 | 11 |
| 2 | 4 | 11 |

The overall stability at the level of the aggregates (with a slight tendency to misrecollect in favour of the centre) hides a marked difference in individual statistics; while 80 percent of those who had selected the centre remembered their first choice accurately, only 33 percent of those who had not selected it did not make a mistake when recollecting.

The last question put to the subject, that identifying the perceived winner, establishes the overwhelming superiority of the centre, a superiority I had expected to appear earlier in the experiment.

Measured in percentage, the final choices were as follows:

| | | |
|---|---|---|
| 0 | 0 | 7 |
| 0 | 88 | 2 |
| 0 | 0 | 2 |

This simple test reveals an equally simple but crucial social mechanism. Why had so many of our respondents not selected the centre in the first place? In answer to that question, nearly all of those who had selected a peripheral location said they had reasoned that the hypothetical "other" with whom they were to be matched would, like themselves, want to avoid doing the obvious, would want to be distinctive. They then realized, after the test, that while their assumption was likely to have been correct, it was an assumption that had no single obvious behavioural translation. The centre integrates, the periphery isolates.

A rational player would have noted that the uniqueness of the centre is in the very language we use to describe the individual elements of the nine-dot configuration. The central dot is the only one that can be described by only *one* word (centre); all others need at least two descriptors (such as upper-right or lower-left).

Similar experiments, those of Schelling (1960) in particular, lead one to believe, however, that if any one of the dots in the display had been clearly singled out from the others by nonlocational characteristics (size, for example), it would have been preferred, irrespective of whether it was in the centre or not. Such experiments show also that a notation so distinguished from the others will tend to be perceived as central to the whole system. All the more so if we wish to express spatially the idea of dominance. As in the architecture of Peking's sacred city, we naturally link centrality and verticality; the dominant is perceived to be both central and high. This mutual attractiveness between centrality and height explains why we move so easily, why we slip so unconsciously, from the notion of geographical centre to the notions of social and cultural dominance, and vice versa. When we speak of extremities in the human body, we mean the feet and the hands, not the head; when we speak of central government, we refer to the more powerful of all the units in a state whether that unit is located roughly at the geographical centre or not.

## CITY CENTRES, VITAL CENTRES, AND MENTAL MAPS

Reflecting on the evolution of the city, Soja (1971) suggests that we can usually distinguish two stages. In its period of mobilization, that corresponding to rapid industrialization and population growth, the city is a conglomerate of people who centre the city on themselves. Their mental maps of the city are exclusively a function of their own place of residence. In a second phase, the city dwellers, according to Soja, tend to locate themselves in the city space as a function of the city's own structures, and the spatial location of self is then made as a function of the city's landmarks, very specifically of those landmarks that identify the city centre. Expanding on this notation, I propose to distinguish three, rather than two, phases. To simplify, let us represent (see Figure 1) a metropolis by a large circle, and its centre by a dot located in the middle; let us use an X to signify an individual's place of residence, and a larger dot to signify the vital centre of an individual's behavioural and perceptual space.

Let us now imagine a French Canadian farmer migrating to Montreal one or two generations ago. Typically he has made the move from country to city either individually or with a small nuclear family; unlike the Malay farmer moving to Kuala Lumpur, he does not bring the village ambiance to the city.[5] To this hypothetical French Canadian farmer, the city centre means little; his relevant behavioural—perceptual space links his place of residence, typically in the eastern suburbs of Montreal, to the village where he has left parents, brothers, sisters, and the parish priest, who incarnates legitimate and familiar authority. Physically he is a resident of a city suburb, but psychologically the city to him is a suburb of the village; he looks outward from the city to the village as in phase a of Figure 1.

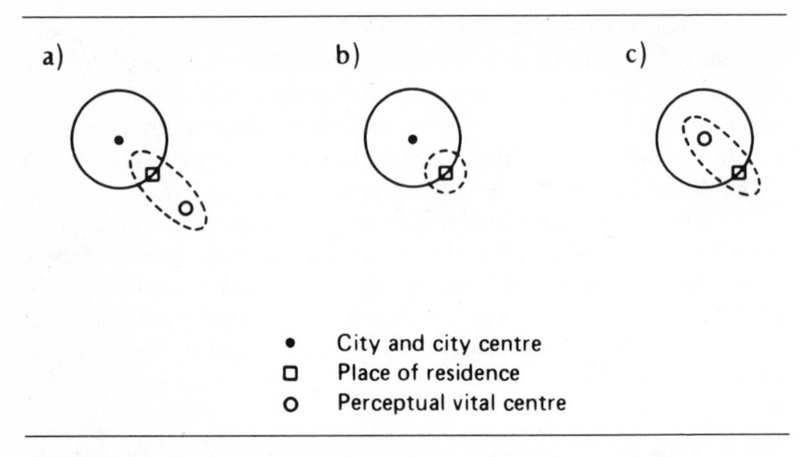


FIGURE 8.1: City centre, place of residence, and vital perceptual centre—three theoretical models

In the next generations, the links with the village will have become tenuous (if only because of the cost of travel and communication), and the stronger links will be mostly with neighbours and coworkers. The relevant behavioural-psychological space is centred on the place of residence—a very small geographical area, a kind of village within the city, but a village more centred than the old village ever was on the family and the home since the public landmarks, churches in particular, are less distinctive and less integrative in the urban setting than in the countryside. This phase is symbolized by phase b of Figure 1.

In the first two phases, the city can easily be tolerant of linguistic diversity since the different linguistic groups occupy different areas of the city and since it matters little to the overwhelming majority of city dwellers what linguistic group "controls" a city centre that is not within their relevant behavioural-perceptual space.[6] Tolerance comes to an end in the third phase (phase c of Figure 1) when the city becomes inward-looking. Even if they remain located at the periphery, the various linguistic groups are now brought into contact and conflict at the city centre, a centre that is increasingly a place of work for white collar workers and also increasingly a privileged place for shopping and entertainment. This third phase of the city's history has some similarities with the third phase of our dot experiment because of the obvious rewards to the concentration of one's attention and behaviour on the centre; the rewards are in the increased freedom of choice and in the reduced communication cost resulting from the concentration of economic and social functions. There is a drawback, however, when central location maximizes unwanted social interaction among linguistic groups that might prefer not to meet. Economic and transport rationality argues for one centre where—in the case of Montreal, at least—ethnic rationality would argue for two. But could the city not develop two different ethnic centres? That is most unlikely when the city has reached the modern phase of its history (phase c). A single illustration of the overwhelming effect of economic and transportation rationality on ethnic segregation will suffice. Up to the 1950s, while Montreal was still by and large in phase b of its evolution, the major downtown department stores operated almost exclusively in English; in prewar days, these stores did not even advertise in the French edition of the telephone book. The francophone equivalents to the downtown Eaton's and Simpson's were the Dupuis Frères stores located in the midst of the francophone residential areas. In the 1960s and 1970s (phase c of the city's evolution)—especially after the opening of the underground metro that gave easy access to the central commercial district—all major downtown stores advertised in French and served francophone clientele; as a result, the Dupuis Frères stores closed their doors in 1977.[7]

The two major Canadian linguistic groups thus came increasingly into contact in Montreal's central business district, in an area that is not only a geographical, communication, and business centre, but also, increasingly, a psychological centre—that by which the modern city is defined. It became a modern city with very distinctive landmarks, both above ground in the form of skyscrapers and underground in the form of a twentieth-century bazaar with shops, restaurants, and entertainment places lining the malls that link metros, hotels, railway stations, and convention centres.

The available data do not enable us to draw a Montreal map of speech, as opposed to a classical map of speakers. (Typically the available data tell us where francophones and anglophones sleep, but not where they speak and in what language.) Should they become available, the maps of speech would probably tell us that French now dominates in the underground central city, but that English is still very much dominant above ground. In the course of its rollback of the English language toward the Ontario border, which began in the nineteenth century, French continues to be stopped at the very edge of Montreal's central business district.

## MONTREAL IN THE CANADIAN LINGUISTIC CONTEXT

More so than religious minorities, linguistic minorities need to achieve and maintain a high degree of geographical concentration in order to ensure their survival. The more frequent the encounters with the language of the dominant group and the more mobile and open the society, the greater the number of individuals and the higher the density needed for a minority language to resist assimilation. The mass and the density needed for linguistic survival in twentieth-century North America is obviously different than what it was in past centuries. An impressionistic guess is that for survival as a full-fledged culture rather than a mere kitchen language, a minority needs now to be so numerous and so concentrated as to be able to sustain a university.[9] French is probably not spoken much more often than Spanish on the North American continent; but unlike Spanish, French has the advantage of being so concentrated as to be able to maintain the universities necessary for its survival. However, the very geographical concentration of French in Canada has caused Montreal, and very specifically Montreal's city centre, to become the linguistic battleground that risks splitting the Canadian confederation.

A hundred years ago, Quebec city, the province's governmental city located in the eastern part of the province of Quebec, was 40 percent English; it is now 96 percent French. And, from the New Brunswick border to the very heart of Montreal, there now extends a zone that is solidly French. In that zone, inhabited by close to five million individuals, roughly 90 percent of the population speak French 90 percent of the time (see Figure 2). The concentration of French in Canada is thus akin to that of French Swiss in Switzerland.[8] But while Quebec city has become overwhelmingly French, Montreal, the province's primate city in the west, while shifting in the last hundred years from an English to a French majority, has not—far from it—eliminated the English presence. The 1971 census indicates that 25 percent of the inhabitants of the greater Montreal area speak English at home, and that these English speakers are concentrated on the west side of the city. The "rollback" of English to the east has been stopped—for more than one generation—at the level of the St. Laurent Boulevard that runs slightly east of Montreal's central business district (see Figure 2). True, as already indicated, French has made remarkable inroads into the city centre, especially underground, but it has thus far failed to gain control of the banking institutions and the head offices of national as well as international companies located above ground. Faced with this obstacle and no longer able to rely on its birth rate—now lower than that of English Canada—the French ethnic group is turning increasingly to political means to ensure its continued social and territorial progression.[10] Linguistic legislation passed by the Quebec National Assembly in 1977 requires that French be the language of work, thus giving employees the right to require that communications from their employers be in French and thus forbidding employers from barring an employee from any position on the sole ground that the prospective em-

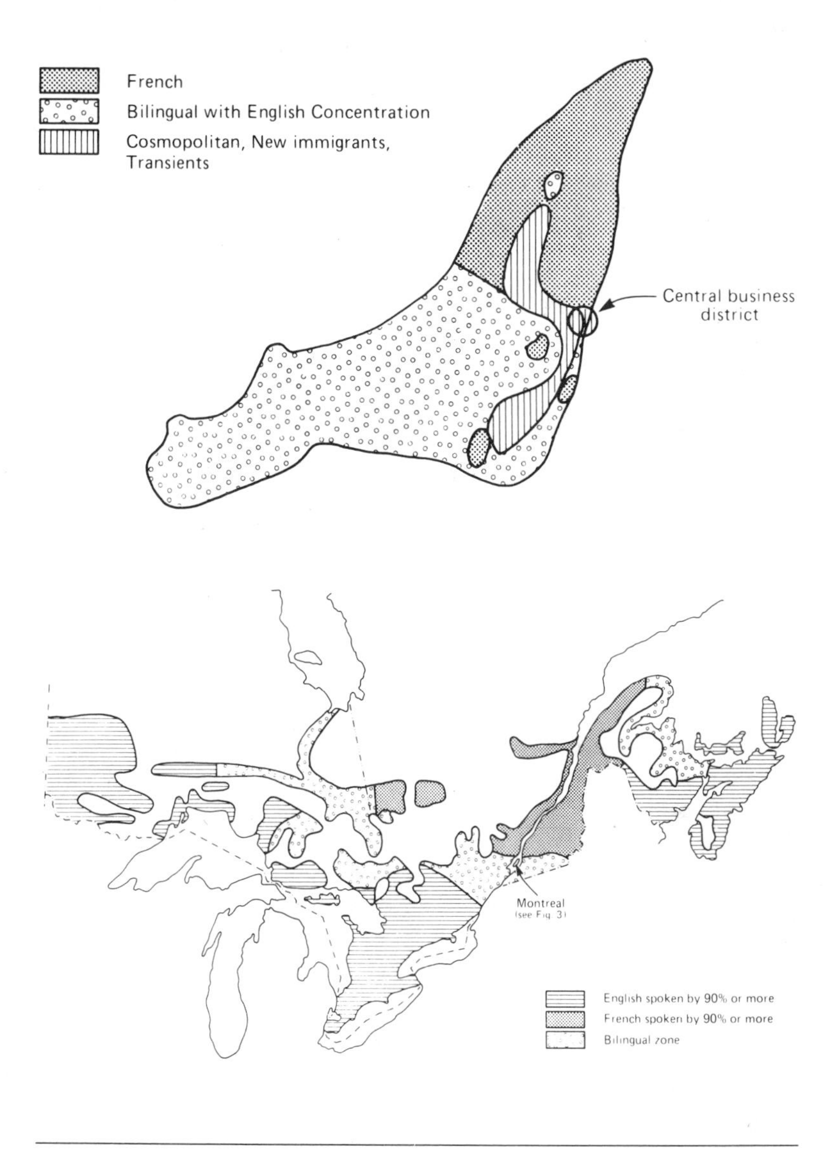


FIGURE 8.2: (Upper) Linguistic zones in Montreal; (Lower) Linguistic zones in Ontario, Quebec, and the Maritimes—areas of concentrated residence
SOURCE: Adapted from Cartwright (1976)

ployee lacks knowledge of a language other than French.[11] The same law requires that commercial advertising and the legal names of firms be in French, a requirement that will have a visible impact on the linguistic "ambiance" of the downtown area. Implementation of these measures will make it increasingly difficult for the anglophones of Montreal to identify with their city. Even more threatening to them is the new requirement that primary and secondary school education be given in French to all newcomers to the province.[12]

## BILINGUAL STRESS AND TERRITORIAL DOMINANCE

It remains to be explained why the central city core, rather than remain English or become French, could not become bilingual. Of course downtown Montreal is already largely bilingual and would remain so even if Quebec were to separate from Canada, but at issue is the overall balance between the two languages; French is now attempting to assume a dominance comparable to that of Dutch over English in Amsterdam, or Finnish over Swedish in Helsinki.

Seemingly paradoxically, the guarantees of French dominance are often sought by educated francophones almost as fluent in English as they are in French. To explain this seeming paradox we need to distinguish the instrumental from the emotional aspects of language use. A language is not a mere tool for communication; it is also a community definer. But the better known a second language, the closer one gets to assimilation into another ethnicity; the more one needs the "other" language as an instrument of social ascension, the more one wishes to protect oneself from assimilation into the "other" group (unless, of course, one wishes to be assimilated). Thus it is always stressful for a minority that wishes to maintain its linguistic identity to have to acquire the language of the dominant group, and it becomes more stressful as that language becomes better known. In addition, it is even more stressful for a dominant group to have to acquire the language of a minority—so stressful that it occurs very rarely.

To reduce the stress caused by bilingualism, to avoid in particular the stress resulting from status inconsistency (not having the same status across valued roles and situations), biligual societies have tended—either naturally or through government regulations—to separate and juxtapose monolingual areas, each assigned to a different language, rather than to mix languages within the same territory. This "Swiss" solution establishes clearly the hierarchy of linguistic dominance: In one zone one language dominates, in another zone it is another language. Consequently the rules of social intercourse become clear: As one crosses the linguistic boundary separating two linguistic groups, the hierarchy between the two languages changes immediately; what was secondary becomes dominant—thus each language has its own areas of territorial security. The Canadian federal government's policy has consisted in resisting the establishment of such boundaries; inversely, the Quebec provincial government has favoured them, first at the level of its own local governments by allowing many of them to operate either in French or in English rather than in both languages (the theoretical norm), and more recently by seeking to make Quebec a monolingual French province within a bilingual Canadian Confederation.[13]

The territorial solution emerging from attempts to reduce the stress (individual and political) occasioned by encounters among francophones and anglophones in Canada and, more specifically, in Montreal appear to point to a Canadian version of the Swiss model: Montreal would become nearly as French as Toronto is English, and these two highly visible primate cities serving and integrating two distinct linguistic communities would be linked politically by a low-profile federal capital. Theoretically, the stress due to bilingualism would not be too disruptive of the Canadian state for Ottawa to be a low-visibility city with a narrow political function. But will the anglophones of Montreal accept marginality in Montreal? Will they accept losing "control" of the downtown area? Will they enter what might be a fourth phase of the city's history in which residents cease to care for the city itself and look outward again, no longer to the village but to a mythical nation, of which they become suburbanites? If the anglophones of Montreal ceased to attach great symbolic value to the linguistic character of the Montreal city centre, if they became increasingly oriented to Canada as a whole, this, undoubtedly, would facilitate the French language overcoming Montreal's central business district in its expansion toward the Ontario border.[14] But, rather than become a kind of Geneva, it is more likely that Montreal will remain—at least in the near future, and unless Quebec separates—a kind of North American Brussels, a continued source of stress and disintegration for surrounding political systems in its inability as a city centre to accommodate two different linguistic groups.

## NOTES

1. An analysis of the use of the notion of centre-periphery by S. Rokkan, J. Galtung, and G. Frank is in McKenzie (1977).

2. For the anecdote of the sages of India telling Alexander how his empire should be governed from the centre, see "Alexander" in *Plutarch's Lives*. The role of centrality as a mental archetype is analyzed in Bird (1977). For the contrast between a "utopian," peripheral world view and a "realist," central-locative world view, see Smith (1971).

3. See Eliade (1957, 1952) and Lévi-Strauss (n.d.: 68–71); for the role of spatial archetypes in shaping political perceptions, see Laponce (1975).

4. In half of the cases the respondents were told that the nine dots symbolized "physical space"; in the other half they were told the display symbolized the "flow of time." This made no statistically significant difference in the pattern of answers.

5. In the 1950s Montreal still attracted mostly young men and women; see Lacoste (1958: 57). The re-creation of small villages within a city is exemplified by the Malays migrating to the cities of Malaya; see Wilson (1967: 42ff).

6. For a factor analysis of census characteristics showing ethnicity to be one of the three major factors explaining location within the city (the other factors being wealth and age of the family), see Wooten (1972).

7. For studies of advertising in the telephone directories of Montreal, see Long (1974) and Bell Telephone Co. of Canada (1938). Both are quoted by Lieberson (1965).

8. About 80,000 students are enrolled at the three major francophone universities: Laval, Montreal, and the Université du Québec.

9. Compared to French in Switzerland, French in Canada has the numerical advantage; it accounts for about 26 percent of the population (measured by language spoken at home), while in Switzerland it accounts for only about 20 percent; furthermore, the French Swiss population is less than two million, thus less than half of the French Canadian population concentrated in Quebec's francophone "security" zone (extending from the centre of Montreal to the east). For historical and contemporary statistics on mother tongues and languages spoken in Canada, see Joy (1972, 1976), Cartwright (1976, and Lieberson (1965, 1970).

10. Downtown businessmen expressed the fear in the mid-1970s that, by concentrating new hotels and convention facilities, the Quebec government might induce a move of the central business district slightly to the east in order to locate it within the French linguistic zone. See the *Financial Post* (1977).

11. An employer can, however, appeal to a government language board to have it established that a given position can legitimately be made subject to the knowledge of a specific foreign language.

12. Under the 1977 legislation, French, at the exclusion of English, became the language of instruction at the primary and secondary levels in all public or state-funded schools. The exceptions were: (1) children of parents who were taught in English in Quebec primary schools, (2) children already taught in English in Quebec, as well as their younger sisters or brothers, and (3) children of parents now residing in Quebec who have received their primary schooling in English outside Quebec, all of whom can be instructed in the English language. This stringent regulation of the language of instruction in schools is not so much intended to induce English speakers already established in Quebec to learn French as it is to force into the French mainstream the new immigrants, especially the non-English/non-French immigrants, who, in the early 1970s, favoured English over French in the proportion of 7 to 3. The text of the 1977 language law (Bill 101) is in *Charte de la langue française,* Assemblée nationale du Québec, 26 aoît 1977. For background to the school and language issue, see Mallea (1977). In 1970, 20 percent of immigrants to Quebec came from French-language countries, 34 percent from English-speaking countries, and 46 percent from other countries; see Gendron Commission (1972: III, 214).

13. For the structure of local governments in the Montreal area, see Bernard (1974). For the language used in local administration before 1977, see Gendron Commission (1972: I, 239). In the 1960s, 61 percent of Quebec's municipalities had obtained the right to operate in only one language: 778 used French, 76 used English.

14. Surveys indicate that, compared to francophones, Quebec's anglophones are markedly more oriented to the federal government. A 1973 comparison of anglophone and francophone law students in Quebec indicates that while 64 percent of francophones consider themselves to be first of all Quebecois, none of the anglophones did so. Of the latter group, 87 percent considered themselves to be Canadians first (Reilly, 1975: 1–2, 67–94). We lack measure of identification with Montreal as a city.

## REFERENCES

Bell Telephone Co. of Canada (1938) Montreal Classified Directory (July).

BERNARD, A. (1974) Profile: Montreal, the Political and Administrative Structures of the Metropolitan Region. Ottawa: Ministry of State.

BIRD, J. (1977) Centrality and Cities. London: Routledge & Kegan Paul.

CARTWRIGHT, D.G. (1976) Language Zones in Canada. Ottawa: Bilingual District Advisory Board.

CHRISTALLER, W. (1966) Central Places in Southern Germany. Englewood Cliffs, NJ: Prentice-Hall.

EISENSTADT, S.N. and S. ROKKAN (eds.) (1973–1974) Building States and Nations, 2 vol. Beverly Hills, CA: Sage.

ELIADE, M. (1957) "Centre du monde, temple, maison," in G. Turri (ed.) Le Symbolisme Cosmique des Monuments Religieux. Rome.

(1952) "Le symbolisme du centre." Revue de Culture Européene: 227–239. Financial Times (1977) December 3.

Gendron Commission (1972) Report of the Commission of Enquiry on the Position of the French Language in Quebec. Quebec: Gouvernement de Québec.

GOTTMANN, J. (1974) The Evolution of Urban Centrality: Orientation for Research. Research Paper No. 8. Oxford: Oxford School of Geography.

JOY, R.J. (1976) "Languages in conflict, Canada 1976." American Review of Canadian Studies 2: 7–21.

(1972) Languages in Conflict. Toronto: McClellan & Stewart.

LACOSTE, N. (1958) Les Caractéristiques Sociales de la Population du Grand Montréal. Montréal: Université de Montréal.

LAPONCE, J. A. (1975) "Spatial archetypes and political perceptions." American Political Science Review 64: 11-20.

LEVI-STRAUSS, C. (1971) Les Mythologiques. Paris: Plon.

LIEBERSON, S. (1970) Language and Ethnic Relations in Canada. New York: John Wiley.

(1965) "Bilingualism in Montreal: a demographic analysis." American Journal of Sociology: 10–25.

LONG, G.L. (1974) Yellow Pages. Montreal.

MALLEA, J. (1977) Quebec's Language Politics: Background and Response. Quebec: Presses de l'Université Laval.

McKENZIE, N. (1977) "Center and periphery: the marriage of two minds." Acta Sociologica: 55–74.

PECJAK, V. (1972) "Affective symbolism of spatial forms in two cultures." International Journal of Psychology 64: 11–20.

ROKKAN, S. (1970) Citizens, Election Parties. Oslo: Universitets Forlaget.

SCHELLING, T. (1960) The Strategy of Conflict. Cambridge, MA: Harvard University Press.

SHILS, E. (1975) Center and Periphery: Essays in Macrosociology. Chicago: University of Chicago Press.
SMITH, J. (1971) "The wobbling pivot." Journal of Religion 51: 135–149.
SOJA, E. (1971) The Political Organization of Space. Resource Paper No. 8. Washington, DC: American Association of Geographers.
WILSON, P. (1967) A Malay Village and Malaysia. New Haven, CT: HRAF Press.
WOOTEN, B.G. (1972) "The urban model, in Montreal Field Guide. Quebec: Presses de l'Université Laval.

# 9

## TERRITORIES, CENTRES, AND PERIPHERIES: TOWARD A GEOETHNIC-GEOECONOMIC-GEOPOLITICAL MODEL OF DIFFERENTIATION WITHIN WESTERN EUROPE

Stein Rokkan

Originally trained in political philosophy and political sociology, I have been increasingly influenced over the last few years by the theoretical revolution in geography. In my early work on the development of mass politics in Norway, I concentrated much of my analytical effort on the deciphering of the marked contrasts between central and peripheral communities and tried to reach some understanding of the modes of interaction between what I called the territorial and the functional dimensions: I studied these processes of interaction as they manifested themselves both in the differential successes of the early waves of political mobilization and in the structuring of regional party systems (Rokkan, 1962). In my later work on the comparative history of mass politics in Western Europe, I combined these two dimensions in an abstract model and tried to show how this model might help to account for variations in the timing of extensions of political rights and in the structure of electoral and organizational alternatives. In the first version of this model I dealt with each territorial polity in isolation from its immediate context: The thrust of the analysis was essentially *typo*logical (Rokkan et al., 1970: Chapter 3). In a later set of articles (Rokkan, 1975, 1973), I tried to recast the model within a *topo*logical framework: I made an effort to locate each case within a geoeconomic-geocultural-geopolitical space and constructed what I called a *conceptual map* of Western Europe. I worked out what amounted to a baseline model of the system of territorial differentiations characteristic of Western Europe and used this model as an engine for the generation of hypotheses about the sources of differences in internal political development case by case. My primary dependent variables were (1) the strength of representative institutions during the phase of absolutist rule from the sixteenth to the nineteenth century, (2) the sequencing of steps toward full-suffrage democracy, (3) the structuring of alternatives for the mass citizenry as expressed in the system of parties, and (4) the vulnerability of each system of mass politics to disruption by monolithic movements during the crisis years from 1918 to 1940. I have not yet been able to present a systematic review of all the hypotheses generated within this framework, and I have done even less to assess in any orderly fashion the wealth of evidence for or against the hypotheses.[1] I am still far from satisfied with the baseline model and want to develop it in further depth before I proceed to a detailed review of its value as an engine for the generation of hypotheses about sources of differences in internal political development.

In this chapter I shall set out, as concretely as I can within the space allotted me, the historical and geographical underpinnings of the basic modelling effort. I shall go beyond the original formulation on one point: I shall discuss the three core components of the baseline model within their *geoethnic context*. The baseline model ignored the concrete ethnic composition of each cell in the geoeconomic-geocultural-geopolitical map of Europe. In this chapter I shall present a first effort to link up a set of geoethnic variables with the geoeconomic-geocultural-geopolitical core components and suggest the contours of an overarching model combining all these elements. This is perhaps a foolhardy enterprise: I am not at all sure that I am on the right track. I publish this early version in the hope that it will inspire others to search for parsimonious ways of systematizing this extraordinary wealth of information. I also entertain the perhaps vain hope that the reactions of my colleagues within political science and geography will force me to recast my model still further and, if possible, make it a more useful tool in the design of comparative analyses across Europe.[2]

## THE CORE ELEMENTS OF THE MODEL

The model is developmental but not unilineal: It brings out the crucial significance of discontinuities, retrenchments, and recombinations of elements. The process of development is analyzed from the vantage point of an isolated primordial community: a closely knit, kinship-regulated local unit covering only a small territory and commanding only elementary technologies of communication. The model posits three part-processes of peripheralization under increasingly powerful systems of long-distance communication and control: one *military-administrative,* one *economic,* and one *cultural*. For each of these processes of territorial aggregation, the model posits a distinctive set of centralizing agencies. These need not control separate physical locations, but may in some cases be found together in close fusion in one dominant centre. Figure 1 spells out this baseline model in further detail.

But the model does not only serve as a tool for the comparative analysis of large-scale efforts of territorial aggregation: It has proved much more directly useful in the study of processes of *fragmentation, retrenchment, and reorganization* of territorial structures. Figure 2 shows how the model can be used to study the combinatorics of processes of breakdown in one concrete case: the disintegration of the western Roman Empire. The midpoints on each of the three core vectors suggest three distinctive modes of disintegration: feudalization, vernacularization, and centre formation on the periphery of the fallen empire. What proved crucial in the Western European case was that these three processes *got out of phase with each other* and that these differences in the timing and impact of the processes produced very different configurations from south to north and from west to east. These contrasts are spelled out in the discussion of the "conceptual map" in sections that follow. The gist of this typological-topological scheme can be stated in two sentences: The emergence of the city belt from south to north in Europe *stopped* the process of feudal fragmentation and produced a new and powerful thrust of long-distance communication and boundary transcendence, while the strengthening of vernacular cultures and the development of major territorial centres at the edges of the empire *accelerated* the break-up of the old system, consolidated new sets of boundaries, and set the stage for the development of a range of highly distinctive political systems within Western Europe. The breakthrough toward merchant capitalism produced a world network of economic transactions and undermined established boundaries, while the emergence of strong nation-states tended to mark off clear-cut boundaries and accentuate territorial identity and citizenship. This is the great paradox of Western European development: The model was designed as a tool for systematic research on the sources and consequences of this paradox.

In this original formulation the model was clearly too abstract: It not only ignored details of social, economic, and political history it also ran roughshod over differences in ethnic legacies and traditional affinities among local and regional cultures within and across the politically and economically

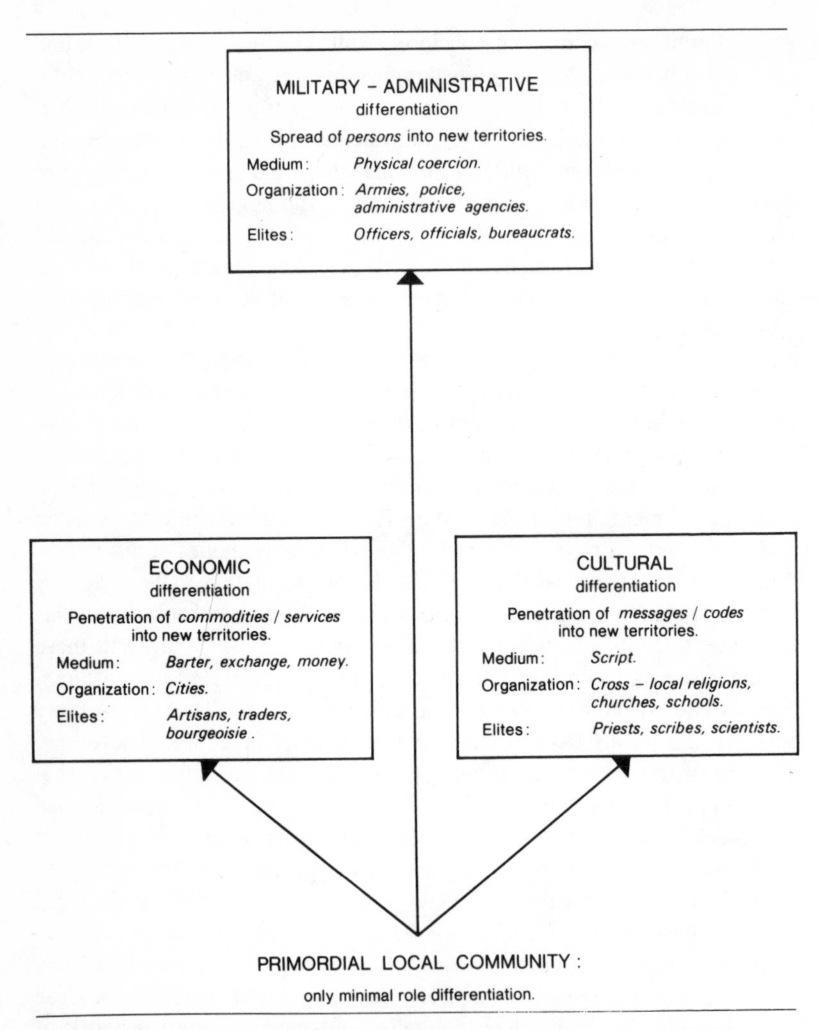


FIGURE 9.1: Three directions of differentiation in large-scale territorial systems

defined boundaries. To put it bluntly: The model left out of account the complex ethnic configurations produced by the successive *Voelkerwanderungen*. The disintegration of Western Europe and the subsequent attempts at territory-building toward the North, West, and East were only too obviously affected by these large-scale movements of ethnically distinctive populations. These differed not only in their languages and customs but also in their ideas of governance, their styles of centre-building, and their resis-

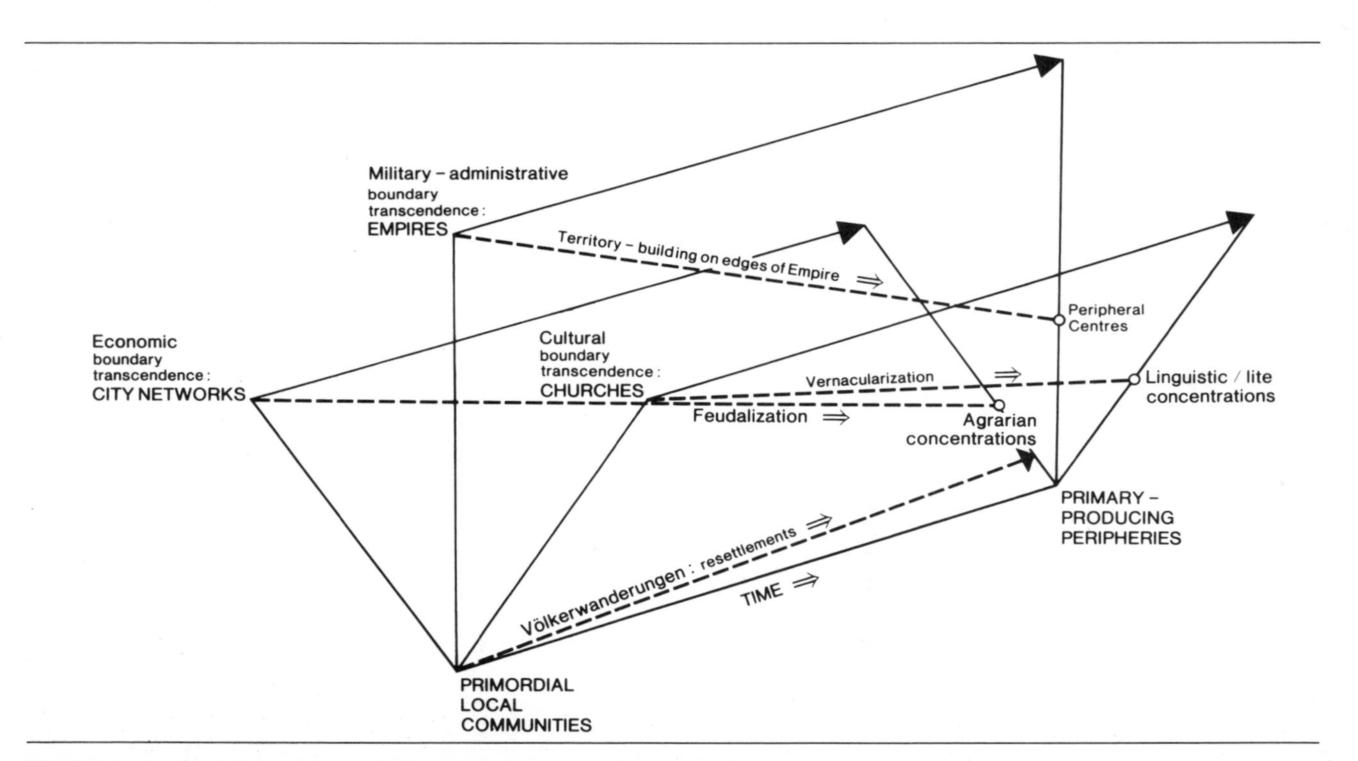


FIGURE 9.2: Chart of the differentiation of processes of disaggregation and reorganization of territorial systems

tance to peripheralization. The time-arrows at the bottom of Figure 2 suggest one set of consequences of these extensive movements of migration and resettlement: Increased integration of peripheral communities within one or more of the three territory-controlling networks, but at the same time increased potential for peripheral protest and for some level of politicization of the peripheral predicament.

On the pages that follow we shall start out our discussion of sources of differentiation within Western Europe with a review of this set of *geoethnic* variables. We are very far from satisfied with the fit between our geoethnic chart and the originally constructed conceptual map, but we hope that this essay will offer some incentive for further efforts of model integration. We are thoroughly convinced that these geoethnic components must be incorporated within a broader model for the explanation of variations across the territories of Western Europe. The importance of efforts on these lines should come out with greatest clarity in our discussion of variations in modes of territorial consolidation and of sources of peripheral protest.

## THE PEOPLING OF WESTERN EUROPE

We cannot get anywhere toward an explanation of the successive changes in the territorial structure of Western Europe without some knowledge of the many waves of migration, conquest, and occupation that have layered the ethnic-linguistic landscape since the Early Iron Age. We can distinguish a total of *seven* major waves. Let us review these in chronological order.

First, the *Celtic* expansion: The Celts moved out from their heartland between the Rhine and the Danube from the sixth century B.C. and occupied large tracts of Gaul, Iberia, Britain, and even Greece.

Second, the long series of *Roman* conquests: The empire moved westward into southern Gaul and Iberia, northward toward the *limes* on the Rhine and the Danube, and then into Britain all the way up to the massive walls built up against the aggressive Picts in Caledonia, what was later called Scotland.

Third, the multiple invasions of the *Germanic* tribes into the crumbling western empire during the fourth and fifth centuries A.D.: The Ostrogoths, Visigoths, and Vandals all covered long stretches of territory on their way and ended up around the Mediterranean; the Lombards settled in northern Italy; the Burgundians in eastern Gaul; the Franks in northern Gaul; and the Jutes, Angles, and Saxons in parts of Britannia.

Fourth, the eighth century wave of *Arab* conquest northward across Iberia and for a brief spell even far into Gaul: This great thrust of Islamic forces was countered by a number of Christian counterthrusts, first into Spain and, later, with the Crusades, across the length of the Mediterranean.

Fifth, the succession of *Viking* raids and conquests: Beyond all the plunder and devastation, these produced lasting settlements in Normandy (911), Ireland (until 1014), England (1013–1042 and again after 1066), and even Sicily and southern Italy (1072–1194).

Sixth, the westward drift of the *Slavs* and the *Finno-Ugric* peoples into the territories to the landward side of the Germans: The most spectacular consequences were the founding of Bohemian, Polish, Hungarian, and Serbian kingdoms during the tenth and eleventh centuries,and the beginnings of a Russian Empire centred on Kiev.

And finally, the *eastward expansion of the Germans* from the twelfth century onwards: This was part of the great drive to Christianize the rest of Europe, but was accompanied by well-planned efforts to colonize and improve poorly used agricultural land. The result was a thorough penetration of German settlers, religious orders, and merchants far into Slavic and Baltic lands, and a long history of conflict between the Marchland rulers and the kingdoms to the east.

These successive waves of conquest and occupation, penetration and retrenchment, produced a complex distribution of ethnic-linguistic groupings across Western Europe. Simplifying in the extreme we can reduce this geoethnic map to the configuration in Table 1.

Starting from the seaward fringe we can distinguish four sets of ethnic groupings along a west-east gradient. First, an *Atlantic periphery* made up of the Celtic and Basque lands and, after the collapse of the early Norwegian North Sea Empire, even West Norway, the Orkneys, Shetland, the Faeroes, and Iceland. Second, the *Western coastal plains*, the heartland of the early seaward kingdoms (the Danish, the Anglo-Saxon, the Frankish, and, considerably later and in a different context, the Iberian). Third, the *central plains* between the Meuse-Rhône line and the Elbe, the heartland of the German-Roman Empire. And fourth, the *landward periphery* caught in the cross-pressure between German and Swedish empire-building thrusts and the resistance of the Slavs, the Magyars, and the Finns.

Each of these four west-east slices can in turn be divided into at least three distinctive layers from north to south. First, the *lands beyond the reach of the Roman Empire:* Ireland, Scotland, northern Germany, Scandinavia, Poland, and the Baltic. Second, the *imperial lands north of the Alps:* England and Wales, France, Switzerland, southern Germany and Austria, and Hungary. Third, the *Mediterranean* lands, the territories most heavily imprinted by Latin institutions and least influenced by the Germanic invaders.

These territorial distributions provided the ethnic-linguistic infrastruc-

TABLE 9.1 A crude geoethnic map of Europe before the high middle ages

| | *Atlantic periphery* | *Coastal plains* | *Central plains and Alpine Territory* | *Landwards marchlands* |
|---|---|---|---|---|
| Beyond the reach of the Roman Empire | Icelanders<br>Faeroese<br>West Norse<br>Celts: Scotland, Ireland | East Norse<br>Danes | Swedes | Finns<br>Balts<br>Prussians<br>Poles, Lithuanians<br>Moravians, Czechs |
| Territory of the northern empire | Celts: Wales<br>Cornwall<br>Brittany | Angles, Saxons<br>Frisians, Jutes<br>West Franks/<br>Gallo-Romans<br>Normans | Germanic Tribes:<br>Burgundians, East Franks<br>Saxons, Thuringians<br>Alemannians, Bavarians<br>Rhaetians | Hungarians<br>Bavarian settlers<br>Tirolians |
| Mediterranean territories | Basques | Occitans<br>Catalans<br>Corsicans<br>Castilians<br>Portuguese | Lombards<br>Italians<br>Sardinians<br>Sicilians | Slovenes<br>Croats<br>Serbs |

tures for the institutional developments of the High Middle Ages: the first steps toward the consolidation of centralized monarchies, the early leagues of cities, and the first consociational structures. In the next round, the distributions of ethnic identities and affinities determined the character and the cost of linguistic standardization within each of these territorial structures. The development of such central standards was accelerated by the invention of printing and the religious conflicts of the Reformation, and it put the peripheries under heavy pressure to accept the norms set by the territorial centres.

## THE COLLAPSE OF THE ROMAN EMPIRE AND THE FORMATION OF NEW CENTRES[3]

To understand these processes of centre-building and peripheralization we have to refer back to Figures 1 and 2. Figure 1 distinguishes three dimensions of differentiation in the development of large-scale territorial systems; Figure 2 specifies the corresponding processes of "retrenchment" to smaller systems covering shorter distances.

At its height the Roman Empire maximized communications and controls in all three dimensions: economic, military-administrative, and cultural. It controlled a vast network of cities around the Mediterranean and at the same time built up a strong centre for the conquest of territories still at a low level of economic development. And, what was equally important, the Roman Empire also became the essential vehicle of penetration into new territories for a major script religion: Christianity.

In this way, the Roman Empire for some time drew strength from all of the three basic processes of differentiation. The three developments reinforced each other for a while, but generated separate organizational structures with resources of their own. The western empire collapsed as a military-administrative structure in the fifth century, but the city network was still there, as was the Roman Church and the tradition of long-distance communication via alphabetic script. The empire broke up as a political system of territorial control, but much of the economic as well as the cultural infrastructure for long-distance communication was left intact and in fact strengthened after the four or five centuries of conflict with Islam, the other empire-building religion of the Mediterranean.

To account for such processes of break-up and reorganization, we posit three processes of periphery build-up within the territory of disintegrating empires: feudalization, vernacularization, and state-building. The location of these processes can be represented graphically at some halfway point along each of the paths of development (see Figure 2). In the history of the territorial structuring of political systems, it is as important to analyze the processes of retrenchment as it is to study the phases of expansion. The system of states emerging in Europe from the twelfth to the twentieth century can only be understood against the background of the Roman inheritance and the reduction in the range and scope of cross-territorial communications that gradually took place in the wake of the fall of Rome.

The rise of feudal structures is a much-discussed theme in comparative history. Within our simple paradigm, what matters is the rise of intermediary power holders controlling resources in the primary economy. This "parochialization" of economic and political power was widespread in the territories of the old empire. But the process was far from uniform throughout Europe: There were few signs of such a process in the territories north of the old Roman territory, and there were also variations depending on the level of agricultural development and the degree of exposure to the onslaughts of nomadic raiders and armies. What proved important in the later development of territorial units was the level of resource generation within such concentrated agrarian structures and the strategies adopted by the owners and controllers of land in their dealings with the urban bourgeoisie and with the military-bureaucratic agencies of state-building. This is the central thrust of Moore's (1966) important work on the development of large-scale territorial units in Europe and Asia.

Developments were much slower on the cultural front: The Roman Church established itself as the central spiritual authority across all of Western Europe and proved able to maintain its two languages, Greek and Latin, as the dominant standards of elite communication for centuries after the fall of the western empire. But, as Goody (1977) has shown in his penetrating analysis of the social consequences of script, there is a fundamental difference between empires built on ideographic communication and empires using alphabets. In China it was possible to keep the literati and the gentry integrated across a vast territory varying greatly in its local vernaculars. The ideographs had no direct relation to speech and could be pronounced in all kinds of ways even though conveying the same message. In the Roman Empire, Greek and Latin were maintained for centuries as vehicles of elite communication across Europe, but the alphabetic script opened up possibilities for the direct expression of vernacular languages. There was already a strong flourishing of such vernacular literatures in the Middle Ages, but the decisive break with the Greek and Latin standards came with the invention of printing and with the Reformation. These developments opened the floodgates for the mass reproduction of messages in vernaculars and set the stage for the establishment of a variety of national standards of communication in an increasingly fragmented Europe. Gutenberg created an essential technology for the building of nations: The mass reproduction of books, tracts, and broadsheets made it possible to reach new strata within each territorial population and at the same time to confine communication within the limits of the particular vernacular. The Reformation reinforced this process in northern Europe. It meant much more than a break with Rome in matters of theological doctrine; it strengthened the distinctiveness of each territorial culture by integrating the priesthood into the administrative machinery of the state and by restricting priests to the confines of the given vernacular. In Hirschman's (1970) terms, the Reformation built up a wall against cultural "exits" into other territories. This wall was not only an important strategy in legitimizing the new territorial state, but in the longer run was also a crucial step in preparing the broader population for the use of "voice" within their system.

This process was not uniform, however. One of the paradoxes of European development is that the strongest and the most durable systems emerged at the *periphery* of the old Empire; the heartlands and the Italian and German territories remained fragmented and dispersed until the nineteenth century.

To reach some understanding of this paradox we have to reason in several steps:

(1) The heartland of the old western empire was studded with cities in a broad trade route belt stretching from the Mediterranean to the east, as well as west of the Alps northward to the Rhine and the Danube.

(2) This "city belt" was at the same time the stronghold of the Roman Catholic Church and had a high density of cathedrals, monasteries, and ecclesiastical principalities.

(3) The very density of established centres within this territory made it difficult to single out any one as superior to all others; there was no geographically-given core area for the development of a strong territorial system.

(4) The resurrection of the Holy Roman Empire under the leadership of the four German tribes did not help to unify the territory. The emperors were prey to shifting electoral alliances, many of them were mere figureheads, and the best and the strongest of them expended their energies in quarrels with the Pope and with the Italian cities.

(5) By contrast, it proved much easier to develop effective core areas at the *edges* of the city-studded territories of the old empire; in these regions, centres could be built up under less competition and could achieve command of the resources in peripheral areas too far from the cities in the central trade belt.

(6) The earliest successes in such efforts of system-building at the edges of the old empire came on the *coastal plains* to *the west and the north,* in France, England, Scandinavia, and later in Spain, in all these cases the dynasties in the core areas were able to command resources from peripheral territories largely beyond the reach of the cities of the central trade belt.

(7) The second wave of successful centre-building took place on the *landward*

*side:* first the Habsburgs, with their core area in Austria; then the eastern march of the German Empire; next the Swedes; and finally, and decisively, the Prussians.

(8) The fragmented middle belt of cities and petty states was the scene of endless onslaughts, countermoves, and efforts of reorganization during the long centuries from Charlemagne to Bismarck. First, the French monarchs gradually took over the old Lotharingian-Burgundian buffer zone from Provence to Flanders and incorporated such typical trade cities as Avignon, Aix, and Lyons. Second, the key cities to the north of the Alps managed to establish a defense league against all comers and gradually built up the *Swiss Confederation;* similar leagues were established along the Rhine and across the Baltic and North Seas, but they never managed to establish themselves as sovereign territorial formations. Third, the Habsburgs made a number of encroachments both on the west and the east of the belt and for some time controlled the crucial territories at the mouth of the Rhine, triggering the next successful effort of consociational confederation: the *United Netherlands*. Finally, in the wake of the French Revolution, Napoleon moved across the middle belt both north and south of the Alps and set in motion a series of efforts of unification that ended with the successes of the Prussians and the Piedmontese in 1870.

## PERIPHERIES AT THE INTERFACE BETWEEN MAJOR CORE TERRITORIES

Let us proceed to discuss these propositions in further detail. The Roman Empire was symbolically reestablished by the Franks in 800, but this construction remained throughout its history a loose federation of kingdoms, principalities, free cities, and ecclesiastical territories. The division of the empire in 843 proved fateful: The western territory was gradually consolidated into a separate kingdom, and the territory in the middle, Lotharingia, was caught in a protracted conflict between the claims of the expanding Frankish kingdom and the different constellations of resource holders within the empire.

For a while it looked as if the eastern boundary of the French kingdom would remain stuck along the Rhône-Saône-Meuse line, but new opportunities opened up with the rapid decline of the internal cohesion of the empire after the extinction of the Hohenstaufen dynasty in the thirteenth century. The constitutional settlement called the Golden Bull (1356) increased the particularist powers of the princes against the Emperor, and the attempts of the Burgundians to create a strong "northern Lotharingia" with its core in what was later called Belgium created increasing tension along the eastern borders of the French territory. The first eastward expansions across the Rhône-Saône line occurred in the South: The Dauphiné came into French hands as early as 1349 and Provence was united to the French crown as an autonomous administrative unit in 1489. The later boundary changes came with the failure of the Burgundian effort to build up a strong state between the French kingdom and the old empire. The Burgundians straddled the Saône and had established great strength both within the French monarchy and within the empire. The western territory, the *Duchy,* was reorganized as the premier peerage of the kingdom, while the eastern territory, the *Franche Comté,* was a highly valued part of the empire. These two territories were brought together under the same dynasty in 1369 and a great effort of state-building got under way. This "Lotharingian" state was composed of the two Burgundies to the south, the Ardennes, the valley of the Meuse, and the Netherlands to the north. The effort collapsed, however, with the great empire-building thrust of the Habsburgs in the fifteenth century. The heiress of the last great duke married the Austrian archduke Maximilian in 1477 and Flanders and Holland, as well as the Franche Comté, came under the sway of the Habsburgs. The sixteenth and the seventeenth centuries saw protracted conflicts between the French and the Austrian-Spanish Habsburgs over these Lotharingian lands. The Thirty Years' War brought Metz and a number of other towns on the northeastern frontier further into French hands. Louis XIV continued these expansionist policies with great vigour and conquered the Franche Comté, the Alsace, and parts of Flanders in 1678, as well as Strasbourg in 1681. Progress was much slower in the Lorraine. After a brief period under the former King of Poland, Stanislaw I, the old duchy was incorporated into the French monarchy in 1766. But this did not stabilize the northeastern frontier for very long. The Lorraine and Flanders were to become the great battlefields between France and the consolidated German Empire in 1870, in 1914–1918, and in 1940, and parts of the Lorraine as well as Alsace changed hands three times over a period of 75 years.

We have gone into such details of territorial history to develop a general thesis: The great efforts of state-building in Western Europe since the High Middle Ages did not only produce a number of subject peripheries in the Celtic-Atlantic West and on the eastern marches, they also generated a number of marginal territories, problem peripheries, *between* the great state-building cores. We could call these *interface* peripheries: They were caught in the cross-fire between dominant centres and were never fully integrated into either of the blocs.

Starting from the North Sea we can identity these buffer territories and problem peripheries between France and different parts of the old empire. First of all, *Flanders* and *Wallonia,* territories straddling the age-old boundary between the Gallo-Roman-Frankish language and various Germanic dialects. Second, *Luxembourg,* a border territory with its own dialect but accepting both French and German as linguistic standards. Third, the *Lor-*

*raine* and the *Alsace*, both heavily Germanic-speaking but strongly oriented toward French culture and French political traditions. Fourth, the *Bernese Jura,* once a part of Burgundy, later a territory under the Prince Bishop of Basle, from 1815 incorporated into a German-speaking canton of the Swiss Confederation. Fifth, the *Savoie,* once a duchy within the Holy Roman Empire, later the core territory of the expanding dynasty that was to unify Italy, but was ceded to Napoleon III as the price for his consent to the formation of the Italian state in 1860. Sixth, the *Val d'Aosta,* a French-speaking enclave within Italian territory. And seventh, the city of *Nice,* a major tourist centre with strong Italian affinities, also ceded to the French by the House of Savoy in 1860.

We could have added to this list the entire *Occitanian* region, the region of the southern Gallo-Roman language, the *Langue d'Oc,* but this clearly could only be classified as an interface periphery by stretching that concept beyond recognition. We *could,* of course, make a case for interpreting the failure of the state builders in Toulouse as a parallel to the Burgundian case, but, first, Occitania was part of the western section of the empire from the start, there was only for a brief period an alternative bloc competing effectively for the control of this territory. Occitania could better be classified, with Scotland, Catalonia, and Bavaria, as "failed-centre" peripheries: territories that *might* have built up their own core structures but were the victims of more effective drives of incorporation from other centres.

We could continue our tour around the entire territory of the old German-Roman Empire and identify a number of further "interface" peripheries. But the character of the cultural and political cross-pressures varied with the blocs confronting each other.

At its height the old empire could count on large chunks of Italy as well as the Germanic lands to the north. The standard literary language of the northern empire was *hochdeutsch;* the standard south of the Alps was increasingly the Florentine variant of Italian. But there remained great variations among dialects and in a few cases even distinctive languages. The most important of these linguistic "isolates" within the old empire were the *Rhaeto-Roman* populations in the *Grisons* in Switzerland and in *Friulia* on the border between Italy and what is now Yugoslavia. We might call these "enclave" peripheries since they do not fit into the dominant culture of the surrounding territory.

By contrast we could classify as "interface" peripheries the territories caught in cross-pressure between two major blocs. Between Italy and the Germanic-speaking territory, the most important such case is the *Alto Adige/South Tirol*. In their efforts to extend their powers across the old empire and beyond, the Habsburgs had acquired Tirol in 1363, Milan and Lombardy in

1535, and Venice and Venetia with the Congress of Vienna in 1815. The territories with clear cut Italian majorities were brought into the unified kingdom in 1859 and in 1866, but there remained an important Italian-speaking minority (as well as a small group of Rhaetians called the Ladins) in what was then called South Tirol, the area stretching south of the Brenner down to Lake Garda. This territory was promised to the Italians in the negotiations that led to their entry into World War I on the side of the Western Alliance in 1915. When the peace treaty was negotiated, the Italians succeeded in imposing a *strategic* boundary along the Alps, but this did not correspond in any way to the *cultural* boundary between German-speakers and Italians. There was a considerable German population south of the Brenner and Mussolini did not succeed in his efforts to "Italianize" the region through accelerated industrialization and public works. The result was a protracted conflict. A new round of negotiations between Italy and Austria took place in 1947 and, after a period of terrorist attacks and counterattacks, a *Proporzpaket,* a detailed package of proportionalizing measures, was signed in 1972. This package holds interesting lessons for similar areas of tension in Western Europe and will be discussed in further detail.

The other interface peripheries to the landward side of the old empire all confront Italians or Germans with Slavs or Hungarians. Starting from the south, these are the major areas of tension:

(1) the *Italian-Slovene* interface in the Trieste-Fiume provinces
(2) the *Austrian-Slovene* interface in Carinthia
(3) the *Austrian-Hungarian* interface in the Burgenland
(4) the extensive *German-Czech* interface in the Sudetenland
(5) the equally extensive *German-Polish* interface, once pushed far to the east toward the Lithuanian border, since World War II placed on the Oder-Neisse line.

We could again add to this list a couple of enclave peripheries: The most important would be the two *Sorbian/Lusatian* territories in the Dresden and Cottbus *Bezirke* of what is today East Germany.

On the northern front of the old empire we can identify two major interfaces: the *German-Danish* in Slesvig and the *Friesian* straddling the Netherlands and Lower Saxony. The Danish kings were active in imperial politics for centuries and held important fiefs south of the linguistic border. These were conquered by the Prussians, but the question of the exact tracing of the military-administrative boundary remained an intractable issue at least until 1920. There were no corresponding conflicts on the German-Dutch border, as the Friesians adjusted peacefully to their predicament.

We can go further north to identify interface peripheries within the extensive territories well beyond the reach of the German-Roman as well as the

Roman Empire. Two cases deserve particular attention:

(1) the *Danish-Swedish* interface in Scania, Halland, and Blekinge
(2) the *Swedish-Finnish* interface in southeast Finland

Sweden conquered the Danish land east of the Sound in 1645 and 1658 and was able to integrate these territories remarkably quickly into their system. One important factor was the establishment of the University of Lund in 1672. This periphery has remained firmly in Swedish hands ever since and there has been no trace of serious irredentism.

By contrast, the Swedish settlement in Finland generated a long history of political conflict. Finland was an early territory of conquest for the kings of Sweden, and the urban elite was predominantly Swedish-speaking until well into the nineteenth century. The extraordinary mobilization of Finnish identity from the 1850s onward set the stage for a series of conflicts, but these were gradually settled through a policy of central equalization and local majority rule for the two language groups.

With the Swedish-Finnish case we have in fact moved into the analysis of a third category of peripheries: We could call these *external* peripheries. We have already reviewed the most important of these in Table 1. The left column in that geoethnic chart lists the Celtic-Atlantic peripheries as "external" to the great state-building endeavours on the coastal plains. We shall revert to these peripheries in our subsequent discussion of unitary-federative structures.

## A CONCEPTUAL MAP OF EUROPE[4]

The long sequences of migration, centre-building, cultural standardization, and boundary imposition produced an extraordinary tangle of territorial structures in Europe: some large, some small, some highly centralized, others made up of differentiated networks of self-reliant cities. The alphabet and the city decided the fate of Europe. The emergence of vernacular standards of communication prepared the ground for the later stages of nation-building at the mass level, and the geography of trade routes made for differences in the resources for state-building between East and West.

The essentials of these differences in the conditions of polity-building can be set out in a conceptual map: a simple two-dimensional typology interpreted within the framework of the overall topology of Europe.

In this map (Table 2) the West-East axis differentiates the economic resource bases of the state-building centres: surpluses from a highly monetized economy in the West; surpluses from agricultural labour in the

East. The North-South axis measures the conditions for rapid cultural integration: the early closing of the borders in the Protestant North; the continued supraterritoriality of the Church in the Catholic South.

This conceptual map reflects the fundamental asymmetry of the geopolitical structure of Europe: the dominant city network of the politically fragmented trade belt from the Mediterranean toward the North, the strength of the cities in the territories consolidated to the seaward side of this belt, and the weakness of the cities in the territories brought together under the strong military centres on the landward Marchland.

The West-East contrast is the underlying dimension in Moore's (1966) analysis. He does not discuss the middle belt, but his contrast between the seaward powers, England and France, and the landward powers, Prussia and Russia, is directly reflected in the West-East gradient in the map. Essentially this was a contrast in the levels of monetization reached at the time of the decisive consolidation of the territorial centres: England and France during the sixteenth century, Prussia and Russia during the seventeenth and eighteenth centuries. In the West, the great surge of commercial activity made it possible for the centre builders to extract resources in easily convertible currency. In the East, the cities were much weaker partners and could not offer the essential resource base for the building of the military machineries of the new centres at the periphery of the old empire. The only alternative partners were the owners of land, and the resources they could offer were food and manpower: crofters, tenants, and smallholders in Sweden; serfs in Austria, Prussia, and Russia. This contrast in the resource bases for political consolidation goes far to explain the difference between the Western and the Eastern systems in their internal structure and in the character of the later transition to mass politics. This is the thrust of Moore's detailed analysis. It does not explain all the cases, however, and does not pinpoint the sources of variations on each side. There are important variations both on the seaward and the landward sides, and these can only be understood through an analysis of the other dimensions of the polities, quite particularly the cultural.

In the conceptual map of Europe, the West-East axis differentiates conditions of *state*-building, the South-North axis the conditions of *nation*-building. In the underlying model of development, the Reformation is interpreted as the first major step toward the definition of territorial nations. Lutherans and Calvinists broke with the supraterritoriality of the Roman Church and merged the ecclesiastical bureaucracies with the secular territorial establishments. This led to a closing of "exit options" on the cultural front and an accentuation of the cultural significance of the borders between territories. The Reformation occurred only a few decades after Gutenberg; the state churches of the Protestant North became major agencies for the

TABLE 9.2 A typology of the political systems of twentieth-century Western Europe

| *System characteristics* | | | | | | | |
|---|---|---|---|---|---|---|---|
| *Size* | *City structure* | *Linguistic structure* | *Seaward peripheries: Sovereign after 1814* | *Seaward nation-states: Retrenched empires* | *City-state Europe: Early consociations or late unification* | *Landward nation-states: Retrenched empires* | *Landward buffers: Sovereign after 1918* |
| Larger | Monocephalic | Integrated | | France (but some resistance among Bretons, Alsatians, Occitans) United Kingdom (Welsh, some Gaelic) | | (DDR) | |
| | | Endoglossic peripheries | | | | | |
| | Polycephalic | Integrated | | | German F. R. and Italy | | |
| | | Endoglossic peripheries | | Spain | | | |
| Smaller | Monocephalic | Integrated | Iceland (Danish never strong) | Denmark | Luxembourg (but exoglossic standards) | Sweden (near polycephalic) Austria | |

| | | | | | |
|---|---|---|---|---|---|
| | Divided | Norway (two standards, one once close to Danish) | | | Finland (one endoglossic, one exoglossic, standard) |
| | | Ireland (Gaelic now very weak) | | | |
| Polycephalic | Integrated | | Portugal | Netherlands (but religious *Verzuiling*) | |
| | Divided: Exoglossic standards | | | Belgium (Flemish, French) Switzerland (German, French, Italian) | (Yugoslavia) |

standardization of national languages and for the socialization of the masses into unified national cultures. In Catholic Europe the Church remained supraterritorial and did not to the same extent prove an agency of nation-building. True enough, the Catholic Church played a major role in the development of peripheral nationalisms in some of the territories of Counter-Reformation Europe, but these were much later developments; they occurred in the aftermath of the French Revolution and took the form of alliances between the Church and nationalist or secessionist leaders against the rulers at the centre, whether Protestant (Belgium before 1830, Ireland from the 1820s onward), Orthodox (Poland, Lithuania), or simply secularizing (the Carlist wars in Spain). Even in the most loyal of the Counter-Reformation states, the churches remained supraterritorial in outlook and never became central agencies of nation-building in the way the Protestant churches did in the North. Whether Protestant or Catholic, the churches had to work with populations varying widely in their openness to efforts of standardization.

The *Voelkerwanderungen* and the struggles of the Middle Ages had produced very different conditions for linguistic unification in the different territories of Europe. The result was a variety of conflicts between claims of territorial control and claims of national identity. There was nowhere a complete fit between the "state" and the "nation," and the conflicts between the two sets of claims were particularly violent in the central trade route belt and in Catholic Europe.

In the northern territories the processes of state-building and nation-building tended to proceed *pari passu,* but even in these systems at the edge of the old empire, claims for territorial control often clashed with claims for separate identities: the English versus the Celts, the Danes versus the Norwegians and the Icelanders, the Swedes versus the Finns. Trade-belt Europe inherited strong linguistic standards from the ancient and medieval empires: Italian in the South, German in the North. But there was no corresponding development at the political level; in these territories national identity came first, political unification only much later. In the "Lotharingian-Burgundian" zone, between the German *Reich* and the French monarchy, the linguistic borders hardly ever followed the territorial frontiers and a variety of developments took place. The Swiss quickly accepted several exoglossic literary standards and never built up a "linguistic nation" of their own; Alsace-Lorraine maintained its Germanic dialect but identified politically with France; Luxembourg also kept its dialect but veered between allegiance to Germany and to France; and Belgium was split between a Flemish-speaking North and a Walloon-speaking South. In the rest of Catholic Europe there were dominant languages at the territorial centres but strong movements of cultural resistance in the peripheries. France went furthest in linguistic standardization but had to keep on "building the nation" in such peripheries as Brittany and Occitania throughout the nineteenth century. The *levée en masse* for the great wars and the nationalization of public education were the decisive vehicles of cultural unification. In Spain the Castilians were never able to build up a unified national culture; the Basque country and Catalonia remained strongholds of regional resistance. Austria tried for centuries to make German the dominant language of Southeast Europe, but the drive never succeeded; with the rise of national ideologies this vast empire soon fell apart into a number of territorial fragments, but even these fragments were still torn by cultural conflicts. The contrast between the Austrian and the Prussian strategies is striking: Austria extended the domain of its state apparatus far beyond the borders of the Germanic language community and acquired a multilingual empire; Prussia had also built up its strength on the eastern marches but in the end turned westward, into the core areas of the ancient German nation. The Catholic power stuck to the supraterritorial idea; the Protestant power endeavoured to acquire territorial control over the one linguistic community. The struggle between *kleindeutsche* and *grossdeutsche* strategies was a struggle over conceptions of the state and of the nation, a struggle between a political and a cultural conception of territorial community.

## FEDERAL VERSUS UNITARY STRUCTURES

We have discussed a variety of sources of diversity within politically defined territories: linguistic minority-majority contrasts, religious divisions, and competitive relations among urban centres. We shall now proceed to review some of the *institutional* expressions of such diversity: the structures of representation, decision-making, and administration set up to coordinate the conflicting interests within the total territory.

Territorial institutions are normally classified on some continuum from loose confederations to highly centralized unitary structures. Riker (1974: 101) defines a federal political organization as one dividing the activities of government between regional units and a central government so as to ensure that each level of government has the final decision in *at least some* of the fields of activity.

Given a definition on these lines, structures will vary along a continuum depending on the ranges and the weights of activities left for final decision at the regional versus the central level.

Such classifications obviously make sense only if the *contents* of the decisions reserved for the lower-level units are specified. The most frequent

| *Alliances* | | *Federations* | | *Fully Centralized Governments* | |
|---|---|---|---|---|---|
| *Weak* | *Close* | *Peripheralized* | *Centralized* | *Regionalized* | *Dictatorial* |
| Organization of American States | EEC | Switzerland Before 1798 | U.S. Yugoslavia | Italy | Hitler's Germany |

solution is to reserve *cultural* matters to the regional units and all matters of *defence* and *foreign policy* to the central unit of government.

In this minimal sense, even *empires* are federative in character. In fact, an often-quoted distinguishing mark of an empire is that it allows autonomy in cultural affairs to a number of subject territories or populations while reserving for the central authority the control of economic transactions and the extraction of surplus resources for military-administrative needs. The *millet* system of the Ottoman Empire is perhaps the most extreme example: The imperial rulers allowed a number of religious communities to keep up their distinctive institutions, languages, and rituals against payment of tribute. There were equally pronounced elements of pluralism and particularism in the German-Roman Empire which was, from the start, a federation of the four original *Staemme,* and the King-Emperor never acquired extensive powers for himself and his court beyond the symbolic and ceremonial ones and the control of appointments, particularly those within the Church. This federal tradition in fact weighed heavily in the history of state-building within the central South-North belt of Europe. Two of the polities within that belt are explicitly federal; one was established as a federation but has moved toward a unitary structure; and two were established as unitary states but have been moving toward new levels of federalization. By contrast, we find hardly any examples of explicit federalization outside this central city belt. Let us look at the cases in turn.

The earliest federal structures emerged *within* the Holy Roman Empire and were essentially alliances of peasant communities or, more frequently, of cities against the encroachments of the stronger dynastic units competing for power within that system. A number of *leagues* were created along the south-north trade routes from the twelfthcentury onward: the Lombard, the Burgundian, the Alsatian, the Swabian, the Rhine Leagues and, of course, the Hanse, the largest and most wide-ranging of them all. All the purely urban leagues lost out in the conflict with territorial princes during the fifteenth and sixteenth centuries. Only two of the defensive alliances survived the great onslaughts of the territorial powers: *the Swiss Confederation* and the *United Netherlands*. Both these structures built up strength at crucial transition points in the trade route system: The Swiss controlled several important passes in the Alps, the Dutch the estuary of the Rhine. These "consociational" polities stood in marked contrast to the unitary nation-states: Switzerland and the United Provinces were the first to practise religious toleration in the wake of the fierce struggles of the Thirty Years' War. But there were important differences: The Dutch territory was much more compact, the Swiss more divided; the Dutch developed a standard language of their own (an "endoglossic" standard), the Swiss did not develop one national language but accepted external ("exoglossic") standards for their three major languages of written communication. These differences in geopolitical position and geocultural structure clearly affected the constitutional reorganizations after the Napoleonic Wars. The Netherlands developed a much more unitary structure; Switzerland retained its federal structure while increasing the powers of the central government, particularly after 1848.

The other territories of the old empire were either incorporated into the stronger nation-states to the west and east or remained a congeries of particularist principalities, cities, and bishoprics until Napoleon showed that unification was not only possible but worth fighting for. There followed two great movements of unification: one within Germany, the other in Italy. Germany was unified in several steps: First a *Zollverein,* a customs union, later, in 1848, a frustrating experience of "consensual" unification under Liberal leadership, and then finally, in 1866, the establishment of the *Norddeutsche Bund* under the military command of the Prussians. Bismarck did not, however, create a unified empire. He was fully aware of the strong particularist loyalties and allowed the principalities and free cities some rights of self-government within a federal structure. This German federalism was reaffirmed in the Weimar Constitution of 1919 and again after the break-up of the old territory between West and East in the Bonn constitution of 1949. Developments in Italy took a different path. There the constitution of the state-building unit, the Kingdom of Sardinia, was extended to the entire territory after the annexations of 1859, 1860, 1866, and 1870. The majority alliance of Piedmontese and Tuscan leaders insisted on the maintenance of a centralized system of government, particularly because of their interest in controlling developments in the southern periphery. The great difference between the German and the Italian unification processes reflected this contrast in the balance between central and peripheral forces. In Germany the highly urbanized western regions were unified from a military periphery: Prussia. In this situation, federalization was the best strategy of unification. In Italy, the unifying power had its base in the highly urbanized North and the state builders did not need to resort to a federalizing strategy to gain control of the southern periphery. Some exceptions had to be made, however, because of strong cultural resistance to centralized rule. The Constitution of 1948 gave separate powers to five autonomous regions: Sicily, Sardinia, and three linguistically divided territories, Valle d'Aosta (French-speaking), Trentino-Alto Adige (large German minority in the South Tirol province of Bozen/Bolzano), and Friulia-Venezia-Giulia (Friulian and Slovene minorities). Steps were also taken to give the other fifteen regions greater powers. From the early 1870s the regional councils took control over the activities of the smallest units of local administration: the communes. There was still considerable tension in the system, however, since there was no fit between the economic structure, centred on the northern triangle of Milan-Turin-Genoa, and the political structure with its centre in Rome.

The Austrian-Italian *Proporzpaket* for the Trentino-Alto-Adige region is of great interest in any comparative study of policies of ethnic accommodation. The solution is consociational rather than federative and it stimulates a series of rights for each linguistic community without establishing clear-cut territorial divisions. In areas of close intermingling of ethnically distinctive populations, this is likely to be the only viable alternative to continued violence.

A roughly parallel solution has been proposed for another interface periphery within the old empire: *Belgium*. The wars of religion divided the Low Countries into a Calvinist North and a Catholic South. The United Provinces of the North were recognized as a sovereign state in 1648, while the South remained subjected to the Habsburg dynasties, first the Spanish, later the Austrian. The Congress of Vienna gave the South to the House of Orange, the rulers of the Northern Netherlands. This merger could not last, however. There was not only a clear-cut religious contrast, but also a marked difference in economic profiles: the South an emerging industrial economy, the North still heavily oriented toward commerce. The French Revolution of July 1830 triggered a revolt in Brussels and the southern provinces succeeded in establishing a sovereign state of their own. But this state straddled the ancient dividing line between Romance and Germanic language in Europe: Francophone dialects to the south, Dutch dialects to the north. The Francophone elites dominated the structure at the time of the liberation, but the nineteenth century saw an accelerating mobilization of the Flemish counterforces. The Catholic Church was able to stem the tide of Flemish separation for some time, but the old system broke down at the end of the 1950s in the wake of a marked upsurge of the Flemish economy. The result was a protracted constitutional conflict. Some form of federalism seemed the obvioius solution but the populations in and around Brussels were so thoroughly intermingled that a simple division of the territory into two equal parts proved impossible. The solution currently under discussion distinguishes between three territories—Flanders, Brussels, and Wallonia—but at the same time allows the citizens to join two dinstinctive cultural communities. This mixture of federalization and communalization is to be made possible by giving citizens within the territory of the capital the right to receive services from and pay taxes to the linguistic community of their choice, irrespective of their exact residence.

The transition from unitary to federal government came earlier and over a longer period of time in the Habsburg lands on the southeastern frontier of the empire. The Habsburg dynasty had been able to build up a major state structure on the marchlands toward the Ottoman Empire and had taken the lead in organizing the Counter-Reformation in the sixteenth century. The Habsburgs were for centuries the most important family of rulers within the German-Roman Empire and, during a brief period in the sixteenth century, came close to establishing complete control across Europe from their strongholds in Austria, Spain, and the Low Countries. But this effort failed and the Habsburgs were definitively reduced to the second rank during the war with Prussia in 1866. The defeat in that year triggered a series of federalizing efforts within the Habsburg territories. The year 1867 saw the establishment of a monarchic union between Austria and Hungary, and the concessions given to the Hungarians only increased the pressure for recognition from other nationalities within the empire. The entire structure collapsed with the defeat in 1918. A number of new states were established within the earlier territories of the Habsburgs, and the German-speaking heartland, Austria proper, was given a federal structure. TheConstitution of 1955 reaffirmed this structure: Austria is composed of nine states *(Laender)* with extensive powers of self-government.

The other states of Western Europe are all unitary in structure, but at least two of them have been taking steps to concede autonomous status to historically distinctive communities brought within their territories through conquest or alliance: the *United Kingdom* and *Spain*.

In military and administrative terms, England was the first territory to be unified within a national framework. The successive invasions by the Romans, the Angles, and the Saxons, and finally the Normans had "flattened" the particularist forces except in the Celtic West and North. The Celtic peripheries were gradually brought under the English crown but were never fully integrated into the political system. Even during the peak period of integration during the nineteenth century, the peoples of these peripheries maintained their separate identities and at least some distinctive institutions. The Anglo-Welsh Union of 1536, the Anglo-Scottish of 1707, and the Anglo-Irish of 1800 did not turn out to be steps in a process of national integration. In fact, these constitutional acts have as frequently been interpreted as possible stepping stones in the development of a federal structure. The forces of resistance grew in intensity in the nineteenth century and reached a first peak with the Irish uprising and the Civil War over the boundary dividing the Republic and the six counties of the North. The next wave of nationalism was triggered by the decline of the economic power of the English centre and the dismantling of the overseas empire.

The boundary drawn between the Republic and Northern Ireland in 1921 had left a time bomb ticking for more than 40 years: The explosion came in 1969 and there is still no end in sight. The cumulation of ethnic, religious, economic, and territorial conflicts within Northern Ireland seems to defy any clear-cut solution. There has been endless talk about "consociational" deals of one sort or another, but the tension built up within each warring camp appears to be so strong as to make it very difficult to devise a *Proporz-paket* of the type worked out for Alto Adige/South Tirol.

A strong nationalist movement gained a number of seats in Scotland in 1974; a weaker one made an impact in Wales and the Labour government was forced, after protracted bickering, to introduce two bills for the "devolution" of a range of powers to Scottish and Welsh Assemblies. Whether these are stepping stones toward a "break-up of Britain" or a fully federal structure is anyone's guess.

The British multinational state was built up through a process of step-by-step incorporation of culturally distinctive territories on the Atlantic periphery. Spain was also built up gradually, but through protracted struggle against a strong enemy, the Muslims to the south. The Spanish state grew out of a coalition of a number of Christian kingdoms fighting the same enemy. Each of these kingdoms were jealous of their distinctive traditions and privileges *(fueros)* and were only willing to join in a loosely structured confederation against the common enemy. Of the four kingdoms, Leon-Castile stood out as the great centralizing force, while Portugal, Navarre, and the crown of Aragon represented the most important fronts of resistance against unification.

On the western periphery Portugal was established as an independent kingdom by the twelfth century and was able to resist all efforts to integrate its territory within an all-Iberian state until 1580. But the union forced on it in that year did not last: Portugal seceded in 1640.

For several centuries the Aragon-Catalonia-Valencia confederacy to the northeast was a stronghold of opposition to Castile. This changed with the marriage of Ferdinand of Aragon to Isabella of Castile in 1469. This was an important step toward the unification of the peninsula, but Aragon, and particularly Catalonia, kept on resisting centralization and on several occasions tried to secede. The Catalans revolted in 1640 and were allied to France until 1652. There followed a period of continued struggle until Catalonia lost its fueros and was integrated into Spain in 1714. Tension between Catalonia and Castile rose to a pitch during the nineteenth century and brought a number of direct confrontations in the twentieth century. With the establishment of the Second Republic in 1931, Catalonia was on the point of declaring itself independent but the Republicans were able to force a compromise by establishing limited home rule under what was called the *Generalidad*. This was abolished by Franco after his victory in the Civil War, but was reconstituted by the Suarez government in 1977.

The Catalan claims for autonomy had been strengthened by the nationalist mobilization of a distinctive linguistic identity: Catalan was historically related to Castilian but had developed a standard and a literature of its own. The resistance of the Basques in Navarra was based on an even stronger sense of separate identity. The Basques had settled in the western Pyrenees long before the arrival of the Romans, the Visigoths, and the Muslims and had never been fully subjected to foreign rule. The Kingdom of Navarre was one of the first to take up the struggle for reconquest of the Iberian territory from the Muslims, and the Castilians later recognized the importance of the Basques as defenders of the northwestern frontier by granting them collective nobility within the realm (Greenwood, 1977).

The Basques never gave up their separate identity, even after centuries of formal integration within the Spanish polity, and again and again rose to defend their fueros against the centralizers in Madrid. The Carlist wars of 1833–1839 and 1872–1876 brought the Basques into alliance with the loyalist Catholics of Navarre and Catalonia against the Castilian modernizers. The tension between traditionalist-nationalists and Castilian centralizers continued in the twentieth century. With the establishment of the Second Republic in 1931, the Basques were promised a Generalidad of the type granted to the Catalans, but the decisions on implementation were delayed until the Civil War and the victory of Franco's forces brought severe repression of all separatist-autonomist movements. A new wave of nationalist mobilization swept the Basque provinces in the 1960s, but the movement was ideologically divided between moderates and extremists and there was little likelihood of a peaceful transition to autonomous status, not least because of the tensions produced by the heavy immigration of Castilian-speakers into the four provinces of Vizcaya, Guipuzcoa, Alava, and Navarra. There were important similarities between the Basque provinces and Catalonia: Both were heavily industrialized and as much oriented to external markets as to the internal Spanish ones. But there were also important differences: The ethnic-linguistic distance between Catalans and Castilians was less pronounced and the tension between Spanish-speakers and

Basque-speakers had been intensified by increased immigration and by the greater strength of the centre in the mass media. There were definite parallels between the conflicts in the Basque provinces and in Northern Ireland. In both cases there exists a marked intermingling of ethnically distinctive populations and a deep division within the nationalist/separatist movement.

Spain had been built up through a slow process of military-administrative unification, but this process at the elite level had not produced a corresponding cultural integration at the mass level. France had also been built up through a process of step-by-step alliances and conquests, but the great Revolution of 1789 and perhaps even more the Napoleonic wars had laid the foundations for a much more determined drive to unify the country culturally as well as administratively.

The Treaty of Verdun had defined a West Frankish kingdom, but this was for a long time as loosely structured as the German-Roman Empire. There was a small royal domain around Paris and Orléans and a complex mosaic of fiefs, some large, some small, across the rest of the territory of the kingdom: the duchies of Burgundy, Gascony, Aquitaine, Brittany, and Normandy; the counties of Champagne, Blois, Toulouse, and Flanders. The decisive breakthrough toward unification and centralization came during the reigns of Philip Augustus (1180–1223) and Louis IX (1226–1270). The royal domain was extended to Normandy and Anjou, and later far into the South with the conquest of Toulouse in the wake of the war against the Albigensian heretics. These acquisitions were threatened during the protracted conflict with the English crown over the control of the West, but at the end of the Hundred Years' War in 1453 the English had lost all their territories on French soil except Calais and the process of consolidation and unification could get under way with renewed force.

But the France of the *ancien régime* remained administratively and juridicially divided. The North of France had retained its customary local laws—this was the *pays de droit coutumier*. South of a line running roughly from the mouth of the Gironde to Geneva, some form of Roman Law prevailed—this was the *pays de droit écrit*. This North-South division was cross-cut by another, essentially centre-periphery contrast. The peripheries had been allowed to keep their estate assemblies for the control of taxation; they were accordingly called *pays d'états*. The central provinces were administrated by officials, *élus*, appointed by the King—these were the *pays d'élection*. The basic policy was to allow greater self-government and lower taxes to the peripheral territories. This reduced the likelihood of changes in allegiance during conflicts at the frontiers of the realm. Artois and Flanders to the north, Lorraine and Alsace in the northeast, the two Burgundies, the Dauphiné and Provence to the east, Corsica, the Languedoc, and Béarn and

Navarre to the south were all important frontier regions given this privileged status to ensure their loyalty in cases of conflict with the Dutch or the Habsburgs. The Breton case was different: This was a periphery on the maritime frontier, incorporated into France first by marriage and later, in 1532, by a formal treaty guaranteeing the traditional local rights.

But these were all pragmatic policies of accommodation—they could not halt the relentless drive for greater centralization. The elites of the constituent territories were increasingly drawn to the King's court, to Versailles and Paris, and this concentration of resource holders made it possible to establish the northern variant of the language, the *langue d'öil,* as the standard for the entire territory. By 1539, Francis I issued an edict proclaiming this variant the sole official language of the realm. The *langue d'oc,* the dialects of the South, remained important in the private lives of the subject populations but lost their institutional base. As late as 1863, officials reckoned that less than half of the territorial population spoke any form of standard French. Outside the provinces between the Loire and the Seine, some form of *patois* prevailed, whether dialects of Romance or distinctive languages such as Breton, Dutch, German, or Catalan.

The distinction between pays d'élection and pays d'états was abolished with the great French Revolution. The entire territory was brought under a unified administration and all citizens were declared equal before the law, whatever their local attachments or identities. In practice, as Alexis de Tocqueville showed in a brilliant analysis, the revolution simply pushed to a further extreme the process of centralization and standardization that got under way in earnest with Francis I. The ancien régime had integrated important sectors of the provincial elites within one territorial system; the revolution went one step further and began a process of integration at the level of the broad masses. The Napoleonic wars and the *levée en masse* began a crucial process of transformation at the level of the peasantry and the working class: the building of a nation within the framework of a military-administrative state apparatus. This was a long process: Eugene Weber has argued that the final transformation of *Peasants into Frenchmen* did not take place until the Third Republic and World War I. Weber lists as the most important forces in this process of integration the universalization of conscription, the introduction of obligatory education, the expansion of the road and railway network, the increased migration into the cities and the new centres of industry, and the spread of the printed word via the press and popular literature. But these forces were strong enough to integrate all the peripheries. Even after the traumatic experience of two world wars there remained important linguistic minorities in Britanny, Flanders, parts of Lorraine and Alsace, Corsica, and Roussillon on the Spanish border. A

TABLE 9.3 Combinatorics of economic, cultural, and politico-administrative diversity

| *City network structure* | *Ethnic linguistic homogeneity* | *Institutional structure* | *Timing of state formation* | *Case* | *particulars* |
|---|---|---|---|---|---|
| Monocephalic | High | Unitary | Medieval | Denmark | German minority in Slesvig |
| M | H | U | Med. | Sweden | City network less centralized; Lappish minority in North |
| M | H | U | 20th ct. | Iceland | |
| M | H | Federal | Med. | Austria | Slovene minority in Carinthia |
| M | Medium | U | Med. | France | Breton, Flemish, Alsatian, Corsican, Occitan, Catalan minorities |
| M | M | U | Med. | U.K. | Federalizing tendencies: N. Ireland, Scotland, Wales |
| M | M | U | 19th ct. | Norway | Two linguistic standards, but no clear-cut ethnic opposition |
| M | M | U | 20th ct. | Ireland | English dominant, but Irish-speakers in West |
| M | Low | U | 20th ct. | Finland | Swedish-speaking minority in Southwest |
| M | L | U→F | 19th ct. | Belgium | Flemish-Walloon opposition; federal-communal constitution under discussion |
| Polycephalic | H | U | 17th ct. | Netherl. | Friesian minority; elements of provincial autonomy |
| P | H | U | Med. | Portugal | |
| P | H | F | 19th ct./split 20th ct. | FRG | Friesian minority; Bavarian particularism |
| P | M | U→F | 19th ct. | Italy | Francophone, Germanic, Friulian minorities |
| P | L | U→F | Med. | Spain | Basque, Catalan minorities; devolved provincial government under discussion |
| P | L | F | Med. | Switzerland | Four linguistic groups |

variety of autonomist movements appeared, particularly in Brittany and Corsica, but none of these has succeeded in bringing about changes in cultural or economic policy. Whatever their particular ideological orientation, the successive French governments have maintained their policies of centralization and standardization. The Gaullist experiment with regionalization was restricted to economic policies and did not in any way imply a move toward a "devolution" of powers from Paris to the provinces: France remained the unitary nation-state par excellence.

We can summarize this brief overview of processes of unification versus federalization in a schematic table (see Table 3). In this we have combined three sets of variables: the first economic, the second cultural, and the third political-administrative. The first dimension in the table groups territories by the *centralization of their city networks:* polycephalic versus monocephalic. The second dimension represents the degree of *ethnic-linguistic unification* of the territory. The third groups the countries by their *institutional policies* from the expressly federative to the most markedly unitary.

There is clearly no direct fit between economic centralization, cultural homogeneity, and institutional structure. In one case, we find a monocephalic city network within a federally structured state; in two other cases, we find polycephalic networks within unitary systems. In one case, Germany, we find a federal structure and a high level of ethnic-linguistic homogeneity within the territorial population; in another, Italy, we find a markedly more heterogeneous population within a near-unitary institutional structure. Practically all the possible combinations have occurred historically, but some of the combinations have clearly made for more peaceful interactions between centres and peripheries than others. Switzerland and Sweden are particularly revealing cases. The Swiss federal structure has no doubt served that multicultural population better than any unitary apparatus. The current conflict within the largest and most "centralist" of the cantons, Berne, seems likely to find a classical solution through territorial division. In the Swedish case, once the Danish and Norwegian territories conquered during the seventeenth century had been culturally integrated, the unitary structure of government certainly served the people well. It is not difficult to see from the review of cases in Table 3 that the greatest strains on territorial equilibria tend to occur in three situations of "poor fit" between economic-cultural distributions and institutional structure: first, in centralized regimes with culturally distinctive populations (e.g., France, Belgium, Finland); second, in centralized regimes with a markedly polycephalic economic network (e.g., Italy and Spain); third, in "mixed strategy" regimes practicising one policy for core areas, another for historically defined communities (e.g., the U.K.).

In the next section, we shall review in further detail the combinations of conditions likely to make for such disruptions of territorial equilibria, whether they have led to successful secessions or simply to protracted mobilizations and countermobilizations in the peripheries.

## A TYPOLOGY OF PERIPHERAL PREDICAMENTS

We have proposed a possible model for the explanation of the formation of territorial *centres* in Western Europe and reviewed differences in territory-building strategies between the powers of the central imperial lands and the nation-states at the edges: *federalizing* strategies predominant within the limits of the old empire, *centralizing* or *mixed core-periphery* strategies outside those limits.

But centres cannot be built up without peripheries, and we cannot complete our discussion of the model without reviewing its consequences for the *classification of peripheral situations*. We suggested earlier a possible typology of peripheries and distinguished between politico-administrative, cultural, and economic peripheries. We also distinguished earlier between external, interface, and enclave peripheries. Let us go back to these distinctions and analyze them in greater detail within the framework of the overall model.

Table 4 highlights the location of peripheries within the conceptual map first set out in Table 2. Each of the territories we have discussed has been placed in the nearest cell within the typological-topological chart. We posit seven types of peripheralized territories from west to east and combine this seaward-landward gradient with an "Arctic-Mediterranean" gradient to produce a further set of peripheries to the north and the south. At each extreme of the west-east gradient we have introduced a distinction between "pure" peripheries and "failed centre" peripheries. In the first category we group the territories that were not unified within any distinctive governmental organization during the Middle Ages; in the second we identify territories with a clear-cut state structure at one point in time but later absorbed into larger formations, whether unitary or federal. Such dichotomies obviously do violence to complex realities: They simply point to a possible gradation to be established against a set of precise criteria. Whatever the precise gradations, the distinction is theoretically of considerable importance. There is a great deal of evidence to suggest that the style and structure of peripheral protest politics varies markedly between territories without any history of distinctive centre-building in the past and territories with a legacy of independent administrative structures at some phase in their earlier history.

The typology of peripheral situations may look excessively complex, but once the basic structure is grasped, it becomes easy to understand the rationale of the total system of distinctions. Table 4 locates the most important peripheries within a grid constituted by three key columns: one for the continuously built-up nation-states on the *seaward side;* one for consociational formations in the old *imperial centre belt;* and one for the continuously developed state formations to the *landward side*[5].

Table 4 identifies two sets of peripheries to the west and east of these three key columns. In both cases we distinguish between territories once unified under a distinctive independent centre, and territories only poorly unified during the Middle Ages. To these four we add three sets of *interface* peripheries: one to the *west* of the old empire, another between the major territories *within* the empire, and a third in the eastern marchlands between Germanic and Slavic-Fenno-Ugric populations.

Table 4 does not give any indications of the latter fate of the different peripheries. It organizes information up to some point in the seventeenth century, but says nothing about any change in the status of the peripheries under the impact of the successive waves of nationalism triggered by the French Revolutions of 1789, 1830, and 1848. Table 5 gives a summary chronology of major events affecting the status of peripheral territories since the Napoleonic wars (see pp. 200–201).

It is easy to see that the greatest number of changes have occurred to the *east* of the Germanic territories. The successive collapses of the Ottoman, the Austrian, and the old Russian Empire created conditions for extensive restructuring of these eastern territories. The decisive steps were taken during and immediately after World War I, but a number of further changes were made, both in boundaries and in constitutional status, in the wake of the Allied victories in 1945. The frontier between Germanic and Slavic territories moved westward and the institution of monolithic Communist rule, even in Roman Catholic and Protestant countries, established an "Iron Curtain" between East and West in Europe. The East-West interface was stabilized for decades, but at the cost of a marked reduction in the freedom of movement across the boundaries.

The main changes in the West took place in the Atlantic periphery: first in Scandinavia, later within the United Kingdom. The wars between France and Germany produced a number of boundary changes at the interface, but no new units were given independent status. The conflict between Francophone and German-speakers in the Bernese Jura brought much bitterness, but seemed likely to be solved under the cumbersome procedure for the divison of cantons prescribed by the Swiss constitution. The 1970s also saw indications of moves toward grants of greater regional autonomy within

TABLE 9.4 Chart for the location of peripheries within the overall territorial structure of a Western Europe

| | *Seaward Peripheries* | | *Coastal Plains* | | *Central Plains/Alps* | | | *Germanic Marchland* | | *Landward peripheries* | |
|---|---|---|---|---|---|---|---|---|---|---|---|
| | 1 | 2 | | 3 | | | 4 | | 5 | 6 | 7 |
| | No. or only incipient, medieval state formation | Distinctive medieval state, but discontinuity | Continuous state formation since Middle Ages | Western interface territories | Continuous state formation since 14th-16th Ct. | Early state formation, later absorbed within larger systems | Internal interface territories | Continuous state formation since 13th-17th Ct. | Eastern interface | Distinctive medieval state, but discontinuity | No or only incipient medieval state formation |
| "Artic" periphery | Greenland<br>Faeroes<br>Shetland<br>Orkneys | Iceland<br>N. Norway<br>↑ | | Jemtland | | | | Norrbotn<br>↑ | | | Lappland |
| Protestant Territory | | Norway<br>Scotland<br>Wales | Denmark<br>England | Bohuslän<br>Scania/<br>Halland/<br>Blekinge<br>Slesvig | | Hanse<br>Germany | Friesia | Sweden<br>Branden-<br>burg/<br>Prussia | Ostro-<br>bothnia<br>Åland<br>E. Prussia | | Finland<br>Baltic<br>Terr. |
| Mixed/ National Catholic Territories | Scottish Highlands<br>Ireland | Brittany | France | Alsace-<br>Lorraine<br>Burgundy<br>Jura<br>Savoie<br>Provence<br>Occitania<br>Corsica | Netherl.<br>Switzerl. | Rhineland<br>Germany<br>Piedmonte/<br>Sardinia | Bohemia/<br>Moravia<br>Bavaria<br>Grisons | → | Sudetenland | Lithvania<br>Poland<br>Serbia | Slovakia |

| | | | | | | | | | | | |
|---|---|---|---|---|---|---|---|---|---|---|---|
| Counter-Reformation Territories | | Galicia<br>Navarre/ Basque Prov. | Spain<br>Portugal | Belgium<br>Catalonia | | Lombardy<br>Papal Territory | Tirol/ Alto Adige<br>Venetia/ Friulia | Austria | Burgenland | Hungary | Slovenia |
| Mediter-ranean periphery | | | Andulucia | Balearic Islands | | Kingdom of Two Sicilies | | | | Croatia<br>Dalmatia | Bosnia<br>Albania |

Belgium, Italy, and Spain, but the details of these reforms have still to be settled.

## FURTHER RESEARCH

This chapter ends in midstream: I have presented a possible model and I have generated a set of typologies of cases and situations, but I have not moved on to any detailed comparative testing of hypotheses. I am currently engaged in a study of conditions for the politicization of territorial identities in Europe and hope to be able to report in detail on data and findings in a collective volume in preparation. We are confident that the distinctions introduced in this essay between external, interface, and failed centre peripheries will prove of immediate value in this effort. This typology obviously needs to be refined and operationalized to be of any direct use in quantitative comparative analysis, but we hope we have already offered enough evidence to suggest its heuristic value.

The distinction between external and interface peripheries expresses a geoethnic dimension: It says something about distances between subjugated and dominant cultures and it helps to differentiate between cases of one-sided dominance and cases of competitive conflict between two dominant cultures (the Basque periphery qualifies as external because of the cultural distance from French and Spanish, but might have been qualified as an interface periphery if the territory on the two sides of the Pyrenees were actively disputed by the two dominant cultures). In our further analysis this geoethnic typology will have to be combined in various ways with geopolitical and geoeconomic classifications. For each chunk of territory we shall have to ask questions not only about its political status (degree of integration under a single territorial authority: subject province versus autonomous region versus constituent member of a federal system), but also about its geoeconomic location. We shall have to ask how advanced the local economy is and what sorts of controls do local-regional elites exercise over transactions with alternative markets for the products of that economy. (A major distinction in our analysis is based on data on the distances between points of production and points of consumption: For primary products this is probably *the* most telling criterion of peripheral status.) Some of these geopolitical-geoeconomic classifications can be generated from the "conceptual map," others will have to be developed and tested within narrower analysis designs.

Let me conclude by offering a brief discussion of one of the key hypotheses in our current work: that the *potential for autonomous peripheral development will increase with economic growth and diversification.* An economically backward periphery is simply not likely to be able to "afford" autonomy. The local leaders are more apt to find the resources they seek through close and loyal cooperation with the authorities in the centre. Claims for greater autonomy are more likely to prove successful if (1) there is some basis of independent economic growth within the periphery and (2) the central authorities find it increasingly difficult to satisfy the demands for subsidies, transfers, and the like made on them by the leaders of the periphery.[6] This is the simplest explanation of the strength of autonomist claims in Catalonia and the Basque provinces. A similar line of explanation has prevailed in analyses of the Flemish-Walloon balance. The applicability of this paradigm to the Scottish and Welsh cases is not quite as obvious. The SNP victory in Scotland came after a period of economic decline. The great wave of protest against the dominance of the centre in London reflected increasing disenchantment with the Union as well as widespread optimism about the prospects of independent growth as a result of the discoveries of oil on the coast of Scotland (*Ed. note:* written in 1978).

The Welsh protest movement has proved weaker, particularly in the most industrialized regions. There may have been disenchantment with the centre, but no corresponding optimism about the potential for independent growth. By accummulating time series for central versus peripheral areas, it should be possible to test alternative hypotheses about the constellations of conditions making for stronger or weaker expressions of peripheral protest.

In other peripheries it appears difficult to substantiate purely economic models for the explanation of the strength of protest movements. The cultural distances between ethnic groups also clearly count. There can be little doubt that the greater distance between the peripheral and the central languages accounts for much of the contrast between the smoother transition to regional autonomy in Catalonia and the much higher level of violence in the four Basque provinces. We have already called attention to the similarity in the Basque and the Northern Irish cases. In both areas there has been a marked accentuation of ethnic contrasts in situations of daily contact between populations of different origins. But in Northern Ireland there is no direct linguistic expression of ethnic identity; instead we find a cumulation of ethnic, religious, and social cleavages. The situation in the Bernese Jura falls somewhere between these two cases. There the strongest wave of separatism came among the Francophone Catholics in the three northern districts (Porrentruy, Délémont, and Franches Montagnes), while the Protestant French-speakers to the south were much more inclined to stick to the *ancien canton* of Berne (the most agonizing situation developed in the city of Moutier in the middle of the Jura, where the Bernese loyalists gained only 68

TABLE 9.5 Schematic Overview of major changes in status of peripheral territories, 1814-1978

| Seaward peripheries | | | *Interface peripheries* | | *Landward peripheries* | |
|---|---|---|---|---|---|---|
| 1 | 2 | 3 | 4 | 5 | 6 | 7 |
| | 1814 Norway: domestic sovereignty | 1815 Jura: incorporated into Berne | 1815 Ticino: separate canton | | | |
| | | 1815 Neuchatel, Geneva, Valais: entered Swiss Confederation as separate cantons | | | | |
| | | 1830 Belgium: indep. | | | 1830 Serbia: indep. under Ottoman suzerainty | |
| | | 1839 Luxembourg: indep. | 1847 Switzerland: *Sonderbund* war | | | |
| | | 1860 Savoie/Nice: integrated into France | | | 1867 Hungary: given equal status in union with Austria | |
| | | 1864 Schleswig-Holstein: province of Prussia | | | Croatia: regional autonomy under Hungary | 1908 Bosniaherzegovenia annexed by Austria |
| | 1905 Norway: indep. | 1871 Alsace-Lorraine: integrated into *Reich* | | | | 1912 Albania: indep. |
| | | | | | 1918 Lituania: indep. | 1917 Finland: indep. |
| | | 1919 Alsace-Lorraine to France | 1919 South Tirol: to Italy | | Poland: indep. | 1918 Estonia: indep. |
| | | 1920 North | | 1920 Åland Island: | Czechoslovakia: | |

| | | | | | | |
|---|---|---|---|---|---|---|
| 1921 Ireland: 26 southern counties indep. | 1936 Basque Provinces: regional autonomy (abolished 1937) | Slesvig: to Denmark<br>1932 Catalonia: regional autonomy<br>1935 Saar: plebiscite, incorporated into *Reich*<br>1939 Spain: victory of Franquists, regional autonomies abolished<br>1940 Alsace-Lorraine: integrated into *Reich* | | special status within Finland<br><br>1920 Dalmatia: transferred from Italy to Kingdom of Serbs, Croats, Slovenes<br><br>Fiume: free city<br><br>Burgenland: plebiscite on boundary corrections | indep.<br>Hungary: indep.<br>Kingdom of Serbs, Croats, Slovenes: indep.<br><br>1938 Sudentenland: annexed by *Reich* | Latvia: indep. |
| 1969 N. Ireland: major wave of violence | 1974 Scotland: SNP electoral gains | 1945 Alsace-Lorraine: returned to France<br>1945 Val D'Aosta: special status within Italy<br>1957 Saar: incorporated into *Bundesrepublik*<br>1974 Jura: referendum on separation<br>1976 Jura: *constituante* for new canton<br>1977 Catalonia: regional autonomy | 1947 South Tirol: renegotiation of status<br><br>1972 S. Tirol/Alto Adige: *Proporzpaket* | 1947 Venetia: agreement on boundary<br>1947 Trieste: free zone<br>1954 Trieste: divided between Italy and Yugoslavia | 1945 Lithuania: incorporated in USSR<br>Poland: major change of boundaries<br>Czechoslovakia: restored<br>Yugoslavia: federation of Serbs, Croats, Slovenes, Bosnians, Montenegrins, and Macedonians | 1945 Estonia and Latvia: incorporated in USSR |

votes more than the separatists in 1974). In the case of the Jura it is difficult to argue for a purely economic explanation of the strength of separatism. At best we could construct a model combining an increase in economic welfare with a greater accentuation of contrasts both among the French and between Francophones and German-speakers.

Perhaps the best strategy in pursuing analyses of this type is to contrast pairs of cases within the overall framework of the unifying model. One contrast that I have always found fascinating is the one between *France* and *Norway*. Weber (1976)[7] has shown how the *Langue d'Oil* standard won out in the *Midi,* first through the incorporation of local elites into the Parisian network during the *ancien régime,* later through the integration of the masses during the Third Republic and World War I. The Occitanian dialects could clearly have offered a basis for a viable counter-language if it had been possible to organize a popular movement in the southern peripheries during the first years of the Third Republic. But there was no mass movement for the defence of a peripheral standard during that crucial phase of mobilization. The Church as well as the network of local notables operating within the Radical party were both organized from the Parisian centre and the peasantry was mobilized into acceptance of the standards of this centre rather than in defence of any counter-standard. The situation was very different in *Norway*. There the distance between the rural dialects and the educated urban standard inherited from the Danes was roughly similar to that between *Langue d'Oil* and *Langue d'Oc* in France (Rokkan, 1967). But in Norway the peasantry defended the periphery and gained recognition for a distinctive counter-standard the *nynorsk*. In the first phase this counter-standard was the work of a peasant autodidact of genius, Ivar Aasen, but the standard he worked out would never have won out if it had not found support in a broad popular movement. The Left *(Venstre)* party in Norway was essentially a coalition of peripheral movements against the dominant urban-bourgeois centre. There was a strong territorial component in the alignments of newly enfranchised citizens against the ancien régime.

Why this difference in the vigour of the peripheral protest movements between France and Norway? Any attempt at an explanation would have to proceed at several levels: the level of the *inherited structure* of territorial communication, the level of the *agencies of mobilization,* and the level of *mass citizenry*. Paris was clearly a much stronger centre than Christiania and it commanded much greater resources both absolutely and relatively. But there was also a difference in the character of the mobilization agencies. In France the successive revolutions had polarized the elite structure and much of politics took the form of competitive drives of mobilization organized from above by the Church, by the secularized apparatus of the state, and by the teachers and the doctors. In Norway there was no such polarization between church and state and the entire centre structure was weaker. The decisive waves of mobilization were organized from outside, through popular movements of different types. The result was a stronger concentration of territorial-cultural oppositions, a genuine conflict between competing standards for the nation. But this very conflict helped to unite the country in the next round: The peripheral standard won recognition and after a few decades gained a strong position within the central culture. In the Norwegian case the peripheral protest movement did not generate any form of separatist action, and the periphery was integrated through a federalizing strategy of accommodation, which was, in fact, a form of "consociation."

## NOTES

1. A brief overview of the model and its primary ramifications was published in a *Festschrift* for Karl W. Deutsch in 1979: *Territories, Nations, Parties*.

2. An expanded version of this text was published in 1979 as part of a "data workbook" on *Centre-Periphery Structures in Western Europe*.

3. This section reproduces a number of passages from Rokkan (1973).

4. This section also reproduces passages from Rokkan (1973).

5. Our emphasis on continuity in the building of independent *states* parallels Seton-Watson's (1977) emphasis on continuities of *nationhood*. Seton-Watson distinguishes "The Old Continuous Nations" (Chapter 2) and the divided territories brought together under the pressures of movements for national unity (Chapter 3). Interestingly his list of "old continuous nations" includes the *Irish*. In our typology this is an ethnically distinctive periphery characterized by a low level of administrative unification during the Middle Ages (the leftmost column in Table 4).

6. Compare the earlier formulation in Rokkan (1970: 121): "Organized resistance against the centralizing apparatus of the mobilizing nation-state appears to have been most likely in three sets of situations:

— heavy concentration of the counter-culture within one clear-cut territory;
— few ties of communication, alliance, and bargaining experience towards the national centre and more towards external centres of cultural or economic influence;
— minimal economic dependence on the political metropolis."

7. See Chapters 6, 17, and 18.

## REFERENCES

GOODY, J. (1977) The Domestication of the Savage Mind. New York: Cambridge University Press.

GREENWOOD, D.J. (1977) "Continuity in change: Spanish Basque ethnicity as a historical process," pp. 81–102 in M.J. Esman (ed.) Ethnic Conflict in the Western World. Ithaca, NY: Cornell University Press.

HIRSCHMAN, A.O. (1970) Exit, Voice and Loyalty. Cambridge, MA: Harvard University Press.

MOORE, B. (1966) Social Origins of Dictatorship and Democracy. Boston: Beacon.
RIKER, W.H. (1974) "Federalism," pp. 93–172 in F.I. Greenstein and N.W. Polsby (eds.) Handbook of Political Science. Reading, MA: Addison-Wesley.
ROKKAN, S. (1975) "Dimensions of state formation and nation-building," pp. 562–600 in C. Tilly (ed.) The Formation of National States in Western Europe. Princeton, NJ: Princeton University Press. (1973)
________(1973) "Cities, states, and nations," Chapter 2 in S.N. Eisenstadt and S. Rokkan (eds.) building States and Nations, Vol. 1. Beverly Hills, CA: Sage.
________(1967) "Geography, religion and social class," in S.M. Lipset and S. Rokkan (eds.) Party Systems and Voter Alignments. New York: Macmillan.
________(1962) "The mobilization of the periphery," in S. Rokkan (ed.) Approaches to the Study of Political Participation. Bergen: Michelsen Institute.
________et al. (1970) Citizens, Elections, Parties. Oslo: Universitetsforlaget.
SETON-WATSON, H. (1977) Nations and States. London: Metheun.
WEBER, E. (1976) Peasants and Frenchmen: The Modernization of Rural France, 1870–1914. Palo Alto, CA: Stanford University Press.

# 10

## THE PERIPHERY AS LOCUS OF INNOVATION

Owen Lattimore

It is rather generally taken for granted that evolutionary developments in the institutions of society begin in the heartland of a high civilization—Rome or China, for example—and radiate outward to its frontiers, where they lag behind developments at the centre in retarded, provincial forms. It is obvious that this is not entirely true: in the United States, in the nineteenth century, capitalistic methods of production often leaped ahead of Europe, because the American society (especially after the liberation of the slaves) did not, as in Europe, have to overcome the vested interests and lingering privileges inherited from a feudal past. In what follows, while not denying the importance of radiation from the centre, I shall try to develop a theory of the alternative, modifying importance of innovation on the periphery, taking as my major example the history of China and its Great Wall frontier.

The year 841 B.C. is accepted as the first fully reliable date in Chinese history, although relative chronologies, based on the succession of rulers, go back much further. By this time the idea of an all-Chinese Empire was well established, but real power, much as under the Holy Roman Empire, passed back and forth among kingdoms that were, in fact, independent. One of the rewards in the pursuit of warfare was control of an Emperor, living in a restricted domain, who was primarily a ceremonial figurehead. Kingdoms were based on regions, each containing a number of cities or towns, walled for defence, each of which was sustained by a surrounding agricultural population. Most trade was, therefore, short range, except for a few special commodities such as salt and iron and, later, silk and tea. Warfare was of a rather static and ponderous kind, and the defence of walled cities was extended to the building of frontier walls, well within China and before the evolution of the Great Wall, dividing the agricultural Chinese from the nomads of Mongolia and Central Asia.

As the agricultural techniques of the Chinese improved, especially in the use of irrigation, several northern kingdoms expanded toward the escarpment of the Inner Mongolian plateau, raising the level of productivity of land that had previously been considered inferior. At the edge of the plateau the expansion halted, and a definite periphery was created, because on the plateau there was not sufficient water for irrigation. South of the periphery the Chinese were becoming, so to speak, more and more Chinese: urban-centred and agriculturally supported. North of the periphery there was much land that could be cultivated, but only in dependence on a rather irregular rainfall, which would have meant an extensive agriculture, in contrast to the increasingly intensive agriculture of the Chinese—much larger acreages, with the use of the plough instead of the heavy Chinese reliance on the hoe and the spade; villages much farther apart and cities even farther apart, requiring a different superstructure of tax collecting, civil authority, and military organisation. In other words, people entering this different geographical zone would have to become "less Chinese" while their kinsmen to the south were becoming "more Chinese."

At the same time, the rainfall of the plateau, often hazardous for agriculture, was enough to support rich pasture. If cultivation was practised at all (and we know that it was, from time to time), it was a wise economic insurance to herd livestock as well, and thus there developed, out of a mixed economy, a full nomadic pastoralism. It is important to stress that this pastoralism was not "primitive," in comparison with agriculture; it was not an activity of people who had not yet learned to cultivate: It was an alternative line of specialisation.

Three interacting phenomena arose out of the confrontation between Chinese agriculture and nomadic pastoralism, all of them innovations and all representing the impact of the periphery on the centre.

First, as the northern kingdoms expanded toward Mongolia, the acquisition of new land and the organising of increased production made them more prosperous and militarily more powerful. Once they confronted the steppe, this expansion halted—as the Roman expansion halted along the Rhine-Danube frontier and the expansion of the British Empire in India halted along the garrisoned line of the north-west frontier with Afghanistan. In all three cases this was because a line of diminishing returns had been reached: A further advance would have thinned out and dissipated the strength of the expanding state; the increased revenue would have been far less than the increased cost. The northern kingdoms began to fight each other more than they fought the barbarians.

Second, in contact with the nomads, the Chinese developed cavalry warfare, especially mounted archers. (In the past, their elite warriors had fought from chariots.) With this new striking force, overwhelming in its time, they did not, however, invade the steppe. They fought each other in order to win the advantage in facing about to conquer the old Chinese heartland, where the greatest wealth and power were to be won. We have specific records of this in the Chinese chronicles of the fourth and third centuries B.C. The final victor in this period of what the Chinese call the Warring States was the northwestern kingdom of Ch'in, which first defeated its northern rivals and then conquered the whole of China down to the Yangtze, and to some extent beyond the Yangtze, creating the first true unified Chinese Empire. It did not last long, but it did set up a permanent model. Because of their cavalry and their mode of fighting, the Ch'in were traditionally considered "half-barbarised" by other Chinese: But they did, from the periphery, create a new kind of China.

Third, prior to this imperial unification each of the northern kingdoms had built its own frontier wall, facing the nomads of the steppe. After the unification, these walls were linked up to form *the* Great Wall. It is essential to grasp the fact that this fortified frontier represented primarily a voluntary Chinese self-limitation rather than the necessity to keep out the "rapacious barbarians." The civilised peoples had writing before the barbarians, and it was a convention of the Chinese historians, as it was of the Greeks and Romans, to attribute to the nomads no motives except a lust for plunder and rapine. This convention must be discounted. In fact, the permanently unsettling factor was the impossibility of establishing mutually satisfactory terms of trade. Civilisation produced much that the barbarians wanted—grain and textiles, as well as luxuries such as silk, but the barbarians did not produce the kinds of things that China (or Rome, or Byzantium) wanted in large quantities. For this reason a barbarian chieftain would threaten the frontier in order to be granted, for political reasons, more trade than would be offered for ordinary commercial profit.

Out of this situation there arose, along the periphery, several interesting phenomena. First, the original North China kingdoms, each with its own stretch of fortified frontier, dealt with one or two separate nomadic tribes. Imperial unification of China and the linking up of a single Great Wall frontier was followed by wars among the nomadic tribes until they, too, had created a unified nomadic empire capable of dealing with this new and greater China.

Second, like the Romans, the Chinese tried to "divide and rule" by dealing with separate elements among the barbarians. At times they were able to engage some of the barbarians as mercenary troops, garrisoned along the frontier. At other times—and this is a phenomenon that has been neglected by the historians—when there was division, confusion, and unrest within China, frontier Chinese looked to barbarian overlords for protection, because they were able to provide more law and order than civilisation itself. In these alternations of power on the peripheries of both China and Rome, whether the civilised people or a barbarian people were in the ascendant, the more powerful a ruler was the more he needed to know how much revenue and how many troops could be raised from each region over which he claimed control. This led to the allocation of domains under hereditary nobles, each owing fixed services to the sovereign. We may, I suggest, be able to ascribe the origin of feudalism to the impact of the periphery on the centre.

# 11

# THE CENTRE-PERIPHERY RELATIONSHIP: PROBLEMS OF SEPARATION IN INDIA, PAKISTAN, AND SRI LANKA

Nirmal Bose

The study of politics—both national and international—cannot be complete without a reference to geography. An interdisciplinary approach to the study of politics is the order of the day. Geography has an effect on regional and national power. According to Mackinder (1904), the father of geopolitics, Eurasia is the "pivot area" and the "heartland" of the world. The peripheral states of these areas of the world will be influenced by the power that will control these pivot areas. The statement may have some elements of exaggeration, but the core of the thesis has not lost all its relevance in international politics even today. Hartshorne (1935) defines political geography as "the study of the State as a characteristic of areas in relation to the other characteristics of areas." Or, in other words, political geography is "the study of political phenomena in their areal context" (Jackson, 1964: 1). Keeping in view the above definitions of political geography, some area studies of South Asia in general and India, Pakistan, Sri Lanka, and Bangladesh in particular are rewarding both from the point of view of power politics of super states and internal dynamics of political developments in each of these states. India, Pakistan, and Sri Lanka are geographically contiguous and share some common historical legacies. At present, in all of these three countries, "political power belongs to persons and groups in the upper class, while the masses remain passive." And further, "changes of political regime have never occurred in response to pressure from the poor masses having become politically aware of their interest and organized for collective action" (Myrdal, 1970: 77). Because of the historical condition, the social and economic conditions of these developing countries are more or less the same, although political development follows a somewhat modernizing course only in India. But even in India, "worse still is the practice, common in all under-developed countries that pretend they are taking radical reform measures, of pronouncing or even legislating large-scale institutional reforms but not implementing them. This breeds cynicism, creates uncertainty about what actually is the established law, and builds up further resistance to implementing the reforms and continuing them" (Myrdal, 1970: 397). Institutional hurdles arising out of mental and attitudinal rigidity partially account for the reasons stated above. In this context, therefore, this chapter proposes to study the centre-periphery relation in its limited scope of a particular geographical area and in an individual political unit. Our objective lies in the very definition of political geography because "The kernel of political geography is the political area. Every pólitical unit describes an areal pattern of nuclear core, constituent or administrative regions, problem areas, vulnerable zones, capitals, strategic spots and boundaries—all affecting success even if not vital to its persistence" (Kasperson and Minghi, 1970: 29).

The peripheral regions in India, Pakistan, and Sri Lanka are not pulling on well with the centres in their respective countries. The trends of secession are quite distinct.

## INDIA

In this discussion of centre-periphery relations, some of the observations of Rajni Kothari are relevant in our introduction. According to Kothari (1970: 89–90), "The partition of India which removed the chief rationale of a loose federation (an earlier idea under which the Hindu and Muslim majority regions would enjoy considerable autonomy) further reinforced the preference for a 'strong centre.' Such a preference, however, went against the strong tradition of a decentralized polity." In the course of his argument, Kothari (1970: 108) notes that there "later occurred another Indian innovation which came from the centre but actually further consolidated the federal character of the Indian polity." This innovation, based on an ancient Indian institution and the Gandhian ideology but for which the main credit should go to Nehru, was a peculiar system of local government from the district downward, known as *panchayati raj* ("democratic decentralization"). This innovation goes well with the nature of Indian society, but the problems with regard to the centre-periphery nexus arise out of the centralized polity.

India is a country with wide cultural diversity. It is very large in size and population. In fact, as of the late 1970s, it was the second most populous and the seventh largest country in the world. It is a country of vastness and variety. There are differences in respect to ethnicity, language, religion, climate, dress, eating habits, and so on. Apart from the centre—where major decisions are made, a number of subcentres at the peripheral level have emerged.

In spite of this great diversity, India emerged as a united composite nation-state because of the protracted and united struggle of the people of almost all the groups for the liberation of the country from alien rule. But after independence (1947), centrifugal tendencies developed in different directions. Nagaland, a periphery on the eastern region, claimed for itself the status of a separate nation, and some people there are still not reconciled to the position of Nagaland as a part of India. Kashmir is enjoying a special status under the constitution of India, and the people of this peripheral region in the North want this to continue, and perhaps they want more. Tamil Nadu, another region (a state) in the deep South, is not inclined to accept domination by the North. The forces of separation and/or secession are still active in these areas.

The reasons for the separatist tendencies are many. First, the language policy followed by the centre is strongly opposed by the people in southern states—Tamil Nadu, Andhra Pradesh, Karnataka, and Kerala. They will not accept Hindi, the official language of the government of India, as their own language, and there have been riots on this question. Second, the complaint of discrimination in respect to economic development is ever present. It is often said that the centre has not done much for the benefit of the people in the southern region. Third, there is a feeling that the government at the centre is always dominated by the people from the North and the West, but not the South.

Demands have been raised in different parts of the country for a suitable change in the constitution to give more powers to the regions—political, economic, financial, and cultural. To absorb the shock, another innovation was the granting of special privileges to the peripheral groups in the society—"scheduled" castes and tribes and "backward classes—in terms of reserved seats and jobs in various legislatures and in the administrative services" (Kothari, 1970: 108). To bring about a harmonious relationship between the centre on the one hand and regional and local units of power on the other, the mechanism of special conference and commissions has been set up. They are often given important powers to help "bridge major differences between the centre and the States." (Austin, 1966). This is the paradoxical adventure of nation-building. The nation is at once built and needs building. The unity is in being and becoming.

## PAKISTAN

Pakistan came into existence in 1947 as a result of partition of India. The people of Pakistan, of both the eastern and western wings, were fully united because of their common religion, i.e., Islam. The Muslims in undivided India felt—and they felt it very strongly—that they were a separate nation and, therefore, they could not remain together in India with other religious groups.

But what happened immediately after the birth of Pakistan? Separatist tendencies started to work. The bond of religion could not resist separatism. In East Pakistan—separated from the western wing of the country by thousands of miles of Indian territory—the people revolted against the centre. They did not accept Urdu, the official language of Pakistan, as their own language and resisted all attempts of interference with their language, i.e., Bangali. They complained that the centre was discriminating against them in the allocation of economic resources. Ultimately in 1971, this peripheral region seceded from Pakistan and emerged as the independent nation of Bangladesh.

The people of other regions, i.e., Sindh, Baluchistan, and the North West Frontier Province, are also fighting for separation in different forms. They feel that they are independent from each other insofar as ethnicity, language, and culture are concerned and, therefore, cannot remain together.

## SRI LANKA

Sri Lanka is a small country, and the people of different ethnic, cultural, and linguistic groups of this country were united for their freedom from alien rule and for their economic development. But cracks developed very soon. The Tamils of Sri Lanka—culturally separated from the rest of the population who are Sinhalese, mostly in the peripheral region of Jaffna—now demand autonomy for their region and equal status for their language, (Tamil) with Sinhalese, the official language of the country. Some of them have also asked for complete secession of their area from Sri Lanka. Recently there were serious riots in the country on these questions. At the back of all this is the complaint of cultural and economic discrimination.

## CONCLUSION

The problems that have arisen out of the centre-periphery relationship in these countries are not just because of lack of communication between the centre and the peripheries, or because of the fact that some peripheral regions are so much at a distance from the centre that they do not and cannot take part in the decision-making process. India is a big country, and it is quite possible that a peripheral region like Tamil Nadu or Nagaland could feel like this. But what about Sri Lanka? Tamils do not live very far away from Colombo—the centre—and there cannot possibly be any communication gap between them from a territorial point of view. The same is true about Sindh or Baluchistan in Pakistan.

The reason for the separatist tendency is mainly cultural. It appears from the in-depth studies of the centre-periphery relationship in these countries that, where there are different and distinct cultural, linguistic, or ethnic groups living together in clusters, they will have a feeling of alienation from the centre. In Bangladesh, there is no such problem because the Bengali people are as one insofar as their cultural identity is concerned, but this cannot be said about Sri Lanka.

It may, therefore, be concluded that whether it is a big country such as India or a small country such as Sri Lanka, if there are separate cultural groups in different peripheral regions, it is better that more power should be given to them. Granting of more power to peripheral regions in these circumstances will not weaken the centre, but rather will strengthen it. They will feel satisfied and there will be no feeling of animosity against the centre. The cooperation with the centre will then be a willing one. Greater autonomy for the regions within the legal framework of the country is the answer for separatist/secessionist tendencies. Federalism is an appropriate remedy. It is not merely a constitutional contrivance designed to hold separate but willing entities together; it is, in a true perspective, a political solution to the problems of a continuous process of modernization. Federation is a multiple relation between the centre and the state, the state and the locality, the state and the region, the region and the locality, and it takes the interlocking process back to the centre. Thus the process is very complex and subtle in any state that has opted for a federal form of government. It is more so the case for India because of the greater diversity of the subcontinent flanked by Pakistan on the northwestern border and Sri Lanka in the south, while the northeastern region is virtually enclosed by China, Bangladesh, and Burma. "Study of the Indian federalising process is even further complicated by the great diversity of the subcontinent which in itself defies analysis" (Franda, 1970). Subregional issues are still unresolved in India even after 30 years of independence—the problems of Telegana in Andhra Pradesh, Belgaum in Karnataka, the hill district of Darjeeling in West Bengal, the tribal areas of Bihar, and the Ao and Angami in Nagaland. Subregional sentiments and urges still remain in Rajasthan, the eastern and western districts of Uttarpradesh and Madhya Pradesh (Kumar and Narain, 1967). A variety of factors—the nature of political leadership, the functioning of the party system, the strength of the pressure groups, factional politics in the state and subregions—all emerging in course of the development of economic life and the consequent modernization of the social system, are to be taken into consideration.

> For realizing national and regional development, seen as mutually consistent goals, there has to be action simultaneously at all levels within the economic and political structure—national, interstate, state and local—and within the public and private sectors of the company. By their very nature, the policies and measures involved place considerable initiative and responsibility for planning and co-ordination in the hands of the governments at the national level. In their turn, state governments and local authorities as well as organizations representing different interests obtain a larger role than they might otherwise have in influencing the dynamic process of economic, social and political change [Singh, 1974: 40].

The interactive process is born of the internal geopolitics of India.

In a broader framework of analysis, not only the politics of India but also that of Pakistan, Sri Lanka, and Bangladesh is determined by the mutual political, economic, and social interactions of the region called South Asia. India is uniquely placed in the region and historically has had the privilege of being the larger and nationally dominant among them all. This may create both empathy and antipathy—empathy, because India's contiguity may release forces initiating economic and social revolution in those countries; antipathy, because of the fear that India's international stature may endanger its political aspirations and social modernization. This antipathy may be strengthened by another factor. A large number of people of Indian origin for various reasons inhabit the regions of those states that are geographically peripheral to Indian state units. These people are loyal citizens of their states of domicile. But a strange feeling of alienation separates the communities of people of Indian origin and the others. These feelings become aggravated when tension grows between these neighbouring countries and India. Illustrations in point are Tamillians in Sri Lanka, Bengali Hindus in Bangladesh,

Hindi-speaking Indians in Pakistan, and Kashmiris in Pakistan-occupied Azad Kashmir.

Federalism is suggested as the solution to the problems posed in this chapter because federalism, with real autonomy of aspirations, seems to be in a position to absorb the shocks resulting from both internal and external compulsions.

## REFERENCES

AUSTIN, G. (1966) The Indian Constitution. London.

FRANDA, M. F. (1970) South Asian Review 3, 3: 199–215.

HARTSHORNE, R. (1935) American Political Science Review 29.

JACKSON, W. A. D. [ed.] (1964) Politics and Geographic Relationships. Englewood Cliffs, NJ: Prentice-Hall.

KASPERSON, R. E. and J. V. MINGHI [eds.] (1970) The Structure of Political Geography. London.

KOTHARI, R. (1970) Politics in India. New Delhi: Orient Longman.

KUMAR, S. and I. NARAIN [eds.] (1967) State Politics In India.

MacKINDER, H.T. (1904) "The geographical pivot of history," Geographical Journal 23.

MYRDAL, G. (1970) The Challenge of World Poverty. Harmondsworth: Penguin.

SINGH, T. (1974) India's Development Experience. New York: Macmillan.

# 12

## ORGANIZING AND REORGANIZING SPACE

Jean Gottmann

The political process develops within the geographical space, and that is the raison d'être of political geography. However, geography does not simply "contain" politics. The political process organizes the space within which it develops, and being a dynamic process, it constantly strives to improve the spatial organization in order to adapt it to change and to fit it to better serve the purposes of government.

Left to the operation of purely natural forces, the geographical environment would evolve slowly. Human action accelerates change. This acceleration has acquired considerable momentum in recent times. The political process has been increasingly concerned with the reorganization of geographical space to answer a rising demand for improvement. Political science and the practical analyses of politicians have, as a result, become preoccupied with the use of space and the improvement of spatial organization. What has been called the "acceleration of history" or the era of "permanent revolution" has also caused politics to give more attention to the factors of time, change, and instability.

Centre and periphery, as a formal geometric relationship, seems to imply an analysis of spatial differentiation on the basis of a certain stability in the geographic distribution of the phenomena considered. The preceding pages, from Strassoldo's Chapter 2 to Bose's Chapter 11, leave, in fact, a strong impression of unsettlement and fluidity. This instability is especially true today, in a period when change and "renewal" are almost the rule. But the frequent references to periods of history also speak of momentous fluctuations in the centre and periphery relationships.

The political process is more in search of an equilibrium, however temporary, than endeavouring to preserve a spatial organization established some time ago. In fact, political stability generally means keeping in power the same people—politicians, parties, or other political groupings. If that aim is achieved by recasting the spatial organization and the regional distribution of weights, such reorganization becomes a means of preserving the political order. The spatial patterns in politics are therefore studied or used as means, not aims.

In special situations, where a carefully balanced political order is contained and constrained by a relatively small closed space, the political process may just aim at the conservation of these circumstances. Isolated states such as Japan during the period of the Tokugawa isolation, or modern Bhutan, can maintain a "frozen political geography" with a well-stabilized centrality and periphery. Situations such as these have been rather exceptional in recent times. At no time in the past could they have been the rule, at least in recorded history.

In his brief illustration of a political "innovation" initiated by a periphery that reorganized a vast empire, Owen Lattimore offers a far-reaching proposition: A region that protests its peripheral status is asking to share in the central functions; it may well attempt to become the centre, to replace the existing structure by another centred on what is, before the reorganization is performed, a dissatisfied periphery. In the Chinese case, a peripheral element reorganized to its own advantage a whole imperial structure. Other great empires of the past have known similar restructuring, as already noted in our introductory Chapter 1. Such renewal of an empire's centrality may or may not involve the transfer of the seat of the central government from the old capital. Whether the political capital moves or not, the main power passes from one group to another. Searching for a new organization of South Asia, Nirmal Bose shows how successive shifts may unsettle balances in a way that adds to the problems of such a vast part of the world. He calls for a new organization of this space on a pluralistic basis to satisfy ideally both the elements in conflict and the need for unity and order.

The examination of centre and periphery as a model in political organization leads to a first conclusion: In the spatial domain the political process consists of frequent shifts in the centre/periphery relationships. Each of the preceding chapters emphasizes some aspects of this dynamism. As hinted at the end of our introductory chapter on the use of this model, it may be better to think of topology rather than geometry as deeply influencing modern political geography.Neither centrality nor periphery seem altogether to disappear or to acquire a diffused pattern; but both move around. With demographic and economic growth, geographical space becomes increasingly partitioned. New nations recently unified, such as Italy or India, still suffer from the many subdivisions affecting their national systems.

The smaller size of political units makes for greater interdependence; self-sufficiency is less and less workable and satisfying in the framework of increased territorial partitioning. This growing interdependence of all the parts of an extremely diversified planet causes the linkages between the parts to multiply. More relationships are established between a greater number of places endowed with centrality functions, some of them narrowly specialized. More relationships are also woven between this greater number of centres and peripheries that continue to both spread and reorganize themselves. A complex web is thus formed, constantly reshaped in some of its sectors, and the whole is animated by flows and currents. Hence the predominating impression of fluidity and instability. Perhaps one ought to begin formulating a *kinetic theory* of political geography.

This inquiry into the concept of centre and periphery has led us to a second stimulating conclusion, which needs to be elaborated further. The kinetic quality of the phenomena discussed here does not seem to result simply from the increase in the quantity of participating elements; the often-mentioned increase in the number of people, governmental units, centres, and peripheral areas is in certain ways instrumental in generating more movement and in expanding and complicating the web, but it is not the fundamental cause.

Rather the essence of the kinetic acceleration that we recognized flows from the wider opening of the whole space concerned to movement and communication. This opening results from the concurrence of many trends: greater freedom of movement of individuals, less stringent terms of employment, more economic specialization and interdependence, technological advances, and also, of course, greater political compartmentalization. The last trend, expressed by the rise of nationalism and regionalism, ought not to obscure the fundamental fact that all this increasing political partitioning developed to satisfy everybody's aspirations to participate more actively and effectively in the global opportunity.

Hence the acceleration and spread of movement, unsettling spatial or hierarchical relationships and reorganizing space. Hence the feeling, repeatedly expressed by Stein Rokkan, of the historical dismantling of the imperial structure Europe inherited from Rome by the multiplication of interrelated centres, old and new. Hence also Raimondo Strassoldo's call for learning to live in a system with many centres, and Lewis Alexander's observations on the decisive role of networks.

In the physical world, relationships of the centre/periphery kind also change, but in most cases so slowly that for any practical human purpose these relationships may be considered to be stable: The orbiting of the planets of our solar system around the sun will continue according to the rules of gravitation between bodies of constant size and density; these are unmovable relationships for all purposes that matter to mankind today. No such stability exists in the domains of politics or economics. All the components of whatever human system that could be considered shift often and irregularly. The spatial distribution of people and political weights evolves in a rather unpredictable fashion. If our inquiry on centre and periphery was intended to clarify the nature of the forces at work, their interplay, and the trends to be projected, we must admit very modest progress only. Most of the concluding statements are more questioning than determining.

Political philosophy, however, has had a long traditon of discussing political stability in the context of centrality and periphery, with rather geographical approaches. In his *Laws,* Plato, advising on the planning of a new *polis,* offers a design aimed at political stability: Most of the population would be settled inland, away from the seashore, to avoid, insofar as possible, maritime and overseas influences. "The territory should be large enough for the maintenance of a certain number of men of modest ambition and no larger" (Laws, 5.737). All necessary relations with the outside world would be dealt with by a small number of specialized civil servants.Plato wanted small, fenced-in states, rather protected from external influence and adventure. Isolation was to help keep the opportunity and the politics stabilized.[1] Such a system could only work if all other states would do likewise and if a general consensus would oppose any interloper.

Plato's assumptions were not realistic. History did not follow his advice. Even his disciple Aristotle began to diverge in his *Politics* toward a more open and broader system, lured by the opportunity then offered to the Greeks. He visualized a strongly united Greece capable of dominating both Asia and Europe. His pupil, Alexander the Great, set out to implement that grand design. However, he quickly realized that his vast empire, which encompassed many great cultures, should not be dominated by Greeks alone, even though all parts of it might be Hellenized. Moreover, the empire was not to be ruled from one central capital, which presumably, in Aristotle's plan, would have been Athens. Alexander planned several great cities, the various Alexandrias, seaports, each of them situated near the mouth of a large river. His empire was to be based on large-scale trade, with the transactional and financial skills well developed by the Greeks, and the whole system resting on a network of large commercial cities. This Alexandrine model emphasizes sea trade, a broadening of opportunity, economic growth in a pluralistic system, with polynuclear networks in a multicultural open space.

This design contrasts sharply with Plato's model. It suggests movement and change, opens the doors to adventure, takes risks with political stability, as Alexander himself soon found out. The debate between the Platonic and Alexandrine views of political organization reflects the opposition between two different philosophies of society, though each feels it must be implemented through a certain organization of space. The size and number of territorial units is less important here than the regulation of accessibility to an open or closed space. This is a fundamental problem linking geography and politics. It is also a basic dilemma in dealing with centre and periphery. For Plato, the periphery must be tightly controlled and subordinated to the centre. Political stability would normally result from this strong centrality. In any case the periphery, in Plato's design, would have no chance to upset the system, which is a closed one. In the open Alexandrine space, the centrality is polynuclear; much more freedom and participation is allowed to various elements within the empire. Peripheral influences could be powerful and unsettling.

Many intermediate solutions can be worked out between the two extreme models. What the Tokugawa shogunate applied to Japan for two centuries closely resembled the Platonic design. What J.G. Fichte suggested for the German states in his *Isolated Commercial State* (1800) seemed inspired by the same doctrine.[2] Fichte wanted to protect the German people from the revolutionary ideas unleashed in 1789 by the French Revolution. In a closed circuit, centrality may remain unchallenged and politics, therefore, rather stable. The American belief, which used to be advertised in New York by large multinational corporations, of "World Peace through World Trade," is a very different, rather Alexandrine approach. It is true in the present era that the interdependence woven between many nations is such that the role of national politics appears less decisive than it has long been. More nations are independent, but the actual significance of the sovereignty allowed to each of them comes into doubt. Rapid change and movement blur the relationships that used to be clearly established, and that blurring affects the role and very nature of centrality.

As we have repeatedly noted, centrality must be located at some place. It is usually a capital city or large metropolis. The number of these centres has recently been increasing rapidly. When hundreds of large cities are sharing in the functions of centrality, the flows along the network considerably deflate the central power vested in each of these cities. However, a relatively small number of them, old and new, retain great authority as centres of important networks, even though some may be specialized and, formally, devoid of political power.

It seems noteworthy that, despite the fluidity permeating our world, many of these very central places are old cities, with centrality established for several centuries at least: Rome, Paris, London, Moscow, Vienna, Peking, Tokyo, Cairo and even New York and Mexico City. Some old capitals, having lost their political role, remain notable economic centres; Constantinople-Istanbul, which was for 1600 years, from Constantine the Great to Kemal Ataturk, a great imperial capital. Other ancient centres have recently been revived: On the basis of cultural and religious roots, political centrality was rather successfully grafted on very old small towns, e.g., Athens or Jerusalem. Amidst all the kinetic momentum of our time, centrality has not become geographically footloose.

This observation leads to a few insights into the present evolution. Historical heritage and cultural differentiation provide essential moorings in the organization of space, as exemplified by many pillars of centrality, as they also provide moorings and arguments in the movement for reorganization by peripheral claims. The "innovating" periphery often prefers to strive for "renaissance" rather than "innovation." Every regionalism offers some cultural arguments in favour of its differentiation: They may be linguistic, social, ethnic, religious, or a mix of these, as illustrated in several of the examples of autonomist or separatist movements reviewed in Stein Rokkan's chapter, or as hinted at in other passages of this book. The roots of regionalism in Italy are also historic, and partly linguistic.

Within the kinetic momentum of this century, most new movements look for some steady base in order to reorganize their sector of space around it, and they usually find these new supports among old symbols that rally the allegiance of their adherents. These symbols are mainly abstract images, often cultural, always evoking deep roots in the past. In my early work on political geography (1950-1952), endeavouring to analyze the processes that led to the political partitioning of geographical space, I suggested the dual model of an interplay between the *movement* factor (in French, *circulation*), which made for change, and the factor of *iconography,* the set of symbols and images that command allegiance.[3] The location of the capital, if it evokes such iconography, does a great deal to establish efficient centrality: e.g., Athens. Even Moscow—a return to the truly Russian past as against the opening of Russia by Peter the Great to Western influences—was a nationalistic symbol Lenin skillfully used to strengthen his new regime. The preservation of iconography could also dictate the closing of national space.

In the kinetic theory of political geography, the interplay of movement and iconography could be used in analyzing the forces working for change or for stability, and in explaining the perception of centre and periphery. Centrality may have become more nomad; it is still very potent, although in an era in which ethical and Platonic ideas have regained popularity, the significance of centrality itself is changing, as it has done before, every time new ideas have emerged about the ways in which society ought to be governed.

Instead of centres, it may perhaps be easier to speak of centrality, meaning a mix of functions all of which have political implications, although many of them are not strictly speaking political; they may be economic, cultural, even narrowly specialized. Central cities work like crossroads, rather than as castles at the top of the hill: They may be described as hinges working within networks.

All these conclusions may leave the reader with a general impression of fuzziness in the pattern of spatial organization. This is due to the need of taking into account the rapid change occurring in so many places and so many respects. It is also due to modern thinking, especially in the Western world, which holds as unethical the enforcement of stability or the preservation of established structures. The organization of space is greatly affected by an acceptance of change and momentum.

The evolution of art in the last century and a half has indeed foretold this trend in social thought and attitudes toward organization of space. As observed by Strassoldo and others, since the great artists of the Renaissance, the emphasis on perspective and on the rules for portraying people and landscapes led to stress on formal organization and centrality. Since the era of the Impressionists, modern art has evolved toward less formal and increasingly unsettled representation of phenomena in space, preferring arrangements of dots and lines, abstract forms, shading of colour, and so on, to the plastically constructed space created by the Renaissance. Architecture and urban planning followed what painters and sculptors had started. Space was used for diffusion rather than convergence. New patterns led to reorganization of the uses of space. Expansion and diffusion apparently avoided centrality.

These new ways of organizing space may make it easier to move people, goods, and even buildings. It allows for greater freedom of action and perhaps more freedom of the individual. A new sociology of art is emerging that is beginning to pay attention to the perception of spatial organization and what it may indicate for the evolution of society.[4] However, our analysis of the political uses of spatial patterns stresses that amidst all the current movement and fluidity, centre and periphery, old formal concepts, are still valid; and that centrality, while challenged, endures and rests increasingly on historically tested symbols that art seems to be prepared to eliminate.

Our time is certainly one of transition, perhaps of a deep and difficult mutation of society. It is penetrating modes of thinking and sets of values. In confronting centre and periphery, we have hoped to advance somewhat the understanding of the political uses of geographical space. While many forms of spatial organization are new, the basic dilemmas are the same, except for the acceleration of the kinetic momentum. Continuing its long evolution, the political process may have to adapt now to a more open and flexible structure of space.

## NOTES

1. I elaborated Plato's search for political stability in my book, *The Significance of Territory* (1973). The Alexandrine model, which I contrast with his, is labelled "cosmopolitan" in some of my previous work, especially in "The Evolution of the Concept of Territory," a paper presented at the IPSA Round Table held in Paris in January 1975.

2. See Note 3 of my introductory Chapter 1, p. 24.

3. The interplay of circulation and iconography was first offered as a model in political geography in my book, *La politique des Etats et leur Géographie* (1952a), and in an article (1952b) summarizing it in English. The interplay of movement and iconography has been discussed in various works on political geography and it acquires perhaps clearer meaning and usefulness in an era of greater momentum of change.

4. This sociology of art was suggested in Focillon's classic book, *La vie des formes* (1939). More recently, the work of Francastel is especially significant, touching upon the representation of space, though without bringing in the present trend to replace geometry by more subjective topology. Some of Paul Klee's writings on modern art are also relevant.

## REFERENCES

FOCILLON, H. (1939) La Vie des Formes. Paris.

FRANCASTEL, P. (1970) Etude des Sociologies de l'Art: Creation Picturale et Société. Paris.

GOTTMANN, J. (1975) "The evolution of the concept of territory." Social Science Information 14, 3/4: 29–47.

________(1973) The Significance of Territory. Charlottesville: University of Virginia Press.

________(1952a) La Politique des Etats et Leur Géographie. Paris: Armand Colin.

________(1952b) "The political partitioning of our world: an attempt at analysis." World Politics 4, 4: 512–519.

## ABOUT THE AUTHORS

LEWIS M. ALEXANDER is Professor of Geography and Head of the Department of Geography and Marine Studies at the University of Rhode Island (Kingston). He is former Director of the Institute of The Law of the Sea and has attended most of the sessions of the U.N. Conference of the Law of the Sea.

NIRMAL BOSE is Professor of Political Science at the University of Calcutta and past President of the Indian Political Science Association.

PAUL CLAVAL is Professor of Geography at the University of Paris—IV and the author of many books on various aspects of geography.

FRANCESCO COMPAGNA is Professor of Economic Geography at the University of Naples and Deputy of Naples, Camera dei Deputati, in the Italian Parliament. He is Editor of *Nord e Sud*, a member of the Trilateral Commission, and has been a member of the Italian government.

JEAN GOTTMANN is Professor of Geography and Head of the School of Geography University of Oxford; he is a Fellow of Hertford College, Oxford, Directeur d'Etudes of the Ecole des Hautes Etudes en Sciences Sociales, Paris, past President of the World Society of Ekistics, and a Fellow of the British Academy.

ALAN K. HENRIKSON is Associate Professor of Diplomatic History at the Fletcher School of Law and Diplomacy, Tufts University, and a Research Fellow of the Woodrow Wilson International Center for Scholars in Washington, D.C.

GEORGE W. HOFFMAN is Professor of Geography and Head of the Department of Geography at the University of Texas, Austin. He has specialized on Southeastern Europe and has served many years on the Joint Slavic Committee of A.C.L.S. and the Advisory Council of the Kennan Institute in Washington, D.C.

JEAN A. LAPONCE is Professor of Political Science at the University of British Columbia and Past President of the International Political Science Association.

OWEN LATTIMORE is Professor Emeritus of Chinese Studies at the University of Leeds, sometime Editor of *Pacific Affairs* (1934–1941) and Director of the Walter Hines Page School of International Affairs at Johns Hopkins University (1939–1953). He is also a member of the American Philosophical Society and the Academy of Sciences of the Mongolian Peoples Republic.

CALOGERO MUSCARA is Professor of Geography at the University of Venice and Director of the Institute of Urbanism, in the Faculty of Architecture at the University of Rome. He is also a member of several commissions in the fields of geography and regional planning.

STEIN ROKKAN was Professor at the University of Bergen and Research Director at the Michelsen Institute (Bergen). He was past President of the International Political Science Association and had specialized on comparative political sociology, theory of elections, and party systems.

RAIMONDO STRASSOLDO is Professor of Political Sociology and Director of the Institute of Poltical Sociology, Gorizia (Italy). He has also been a participant in many international conferences on boundary systems and political sociology.

Mechanic Mike's Machines

# Trains

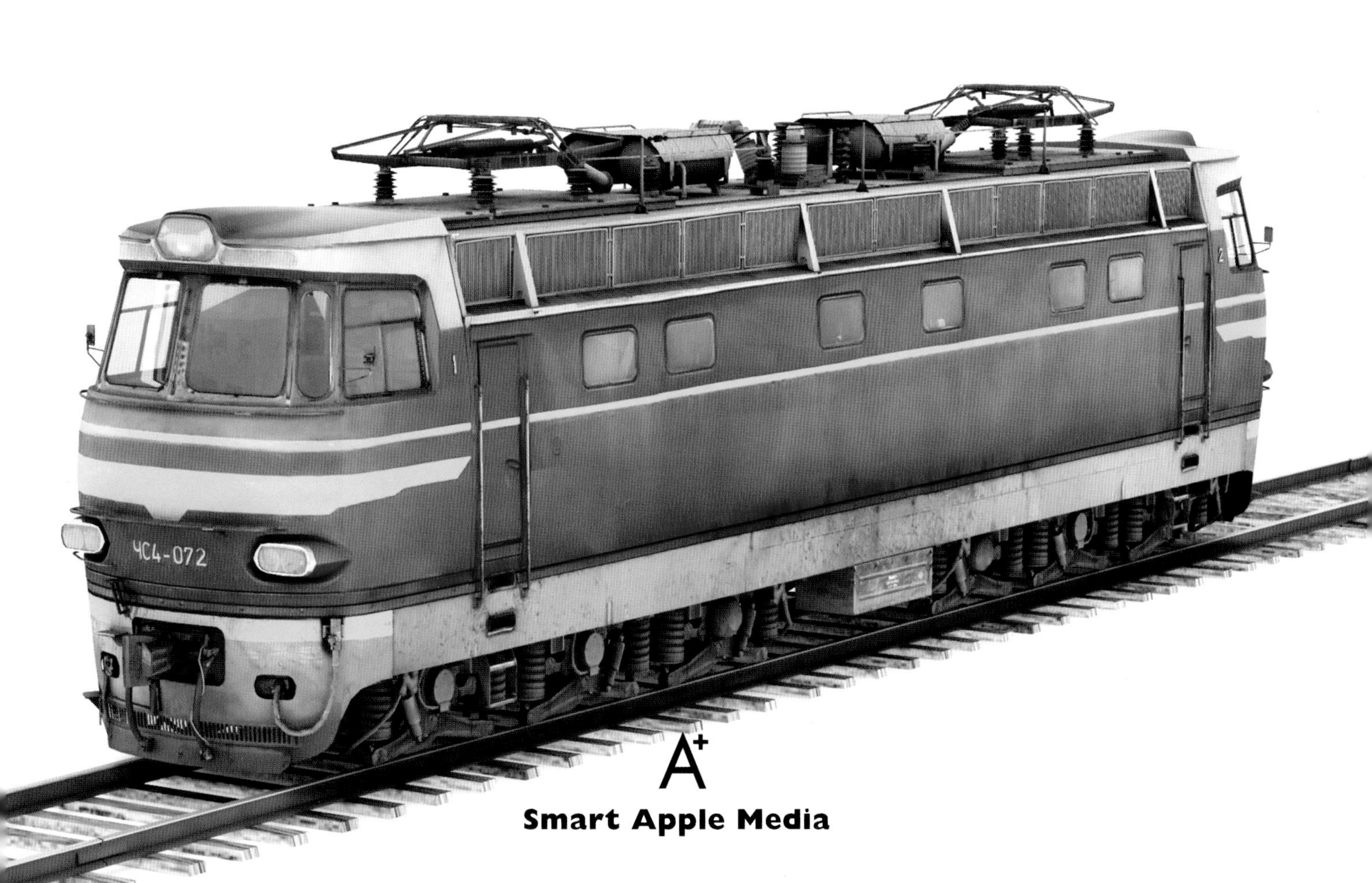


Smart Apple Media

Published by Smart Apple Media, an imprint of Black Rabbit Books
P.O. Box 3263, Mankato, Minnesota 56002
www.blackrabbitbooks.com



Produced by David West Children's Books
6 Princeton Court, 55 Felsham Road, London SW15 1AZ

Designed and illustrated by David West



Cataloging-in-Publication data is available from the Library of Congress.
West, David, 1956- author.
Trains / David West.
pages cm – (Mechanic Mike's machines)
Summary: "This new title introduces young readers to trains. Starting with the original steam powered locomotive, this title teaches young readers all about different trains that have developed. Large-scale pictures accompany a description of each train and are complemented by sidebars containing interesting facts"–Provided by publisher.
Includes index.
Audience: K-3
ISBN 978-1-62588-060-4 (library binding)
ISBN 978-1-62588-099-4 (paperback)
1. Railroad trains–Juvenile literature. I. Title. II. Series: West, David, 1956- Mechanic Mike's machines.
TF148.W469 2015
625.2–dc 3

2013031985

Printed in China
CPSIA compliance information: DWCB14CP
010114

9 8 7 6 5 4 3 2 1

**Mechanic Mike says:**
This little guy will tell you something more about the machine.

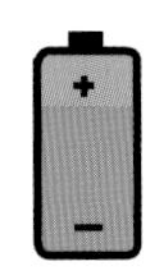

Find out what type of engine drives the machine.

Discover something you didn't know.

Is it fast or slow? Top speeds are given here.

How many crew or people does it carry?

Get your amazing fact here!

# Contents

Early American steam **locomotives** like the "General," built in 1855, had a tall chimney stack and a cow catcher on the front. This was to clear the rail line of anything that might derail the train.

Early steam trains, such as the "General," could reach 20 mph (32.1 km/h) in short bursts.

Steam trains usually had a crew of at least two people. The driver and the fireman who kept the fire in the boiler fed with coal.

Did you know the "General" is famous for its part in the "Great Locomotive Chase" during the Civil War? Union raiders stole the train and were chased by Confederates in other trains.

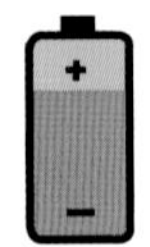

It has a steam engine.

# Steam

The earliest locomotives were powered by steam. Coal or wood was burned to heat up water in a boiler to make steam.

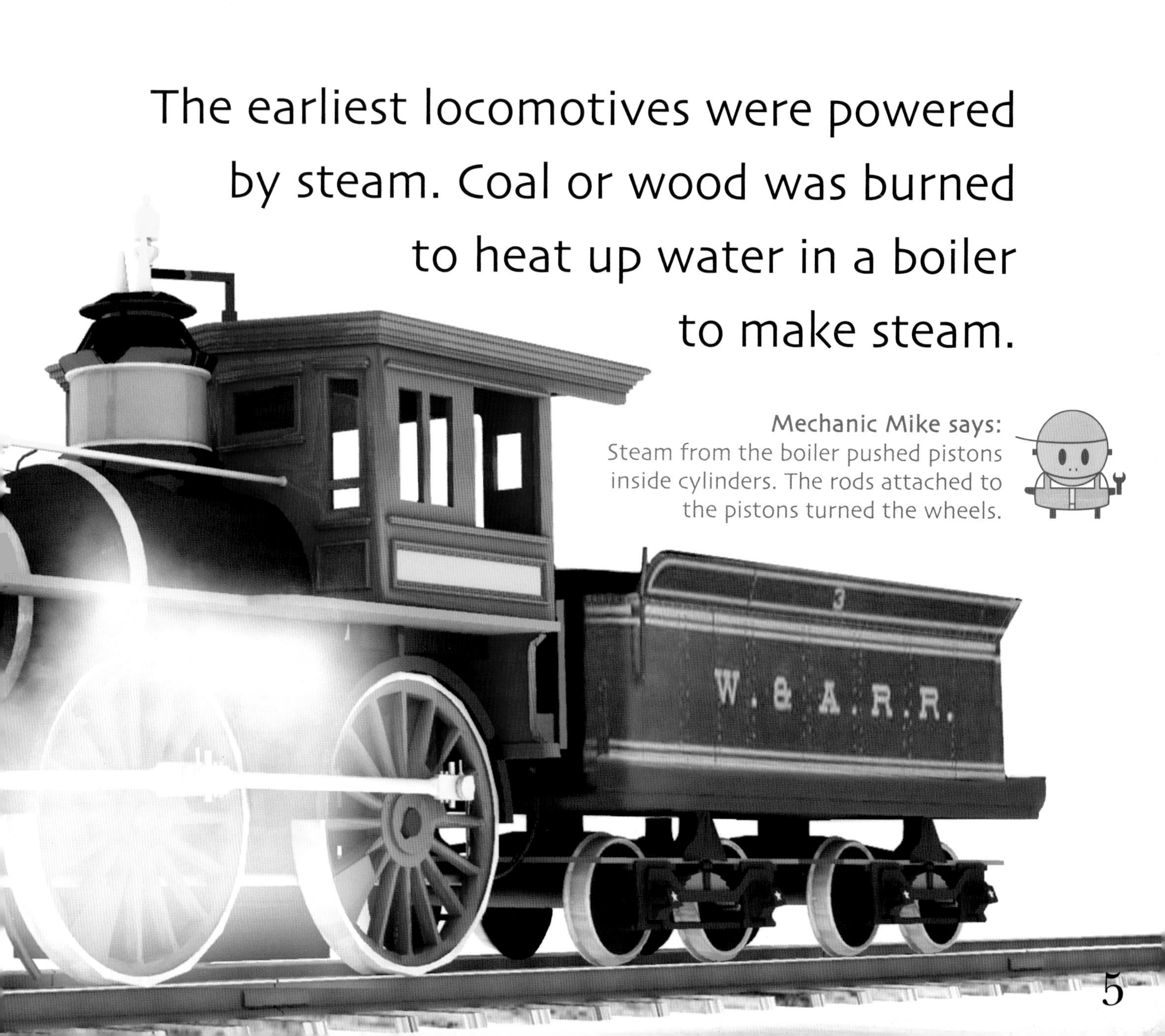


**Mechanic Mike says:**
Steam from the boiler pushed pistons inside cylinders. The rods attached to the pistons turned the wheels.

The biggest **diesels** ever built were the Union Pacific Centennial locomotives. They were 98 feet, 5 inches (30 m) long.

Diesel trains can be fast and powerful. The high-speed diesel train, InterCity 125, used for passenger service in the UK is the fastest diesel train at 144 mph (231 km/h).

A former Santa Fe F45, in Montana, has been converted into a lodge where up to four people can stay.

This diesel-electric EMD F45 has a crew of two.

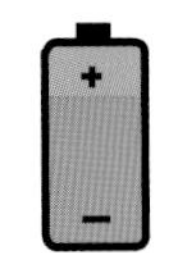

This EMD F45 has a 20 cylinder diesel engine powering electric motors.

**Mechanic Mike says:**
Diesel trains that power electric motors are sometimes called diesel-electric trains.

# Diesel

Diesel locomotives are powered by diesel engines. In some cases the engine powers electric motors which turn the wheels.

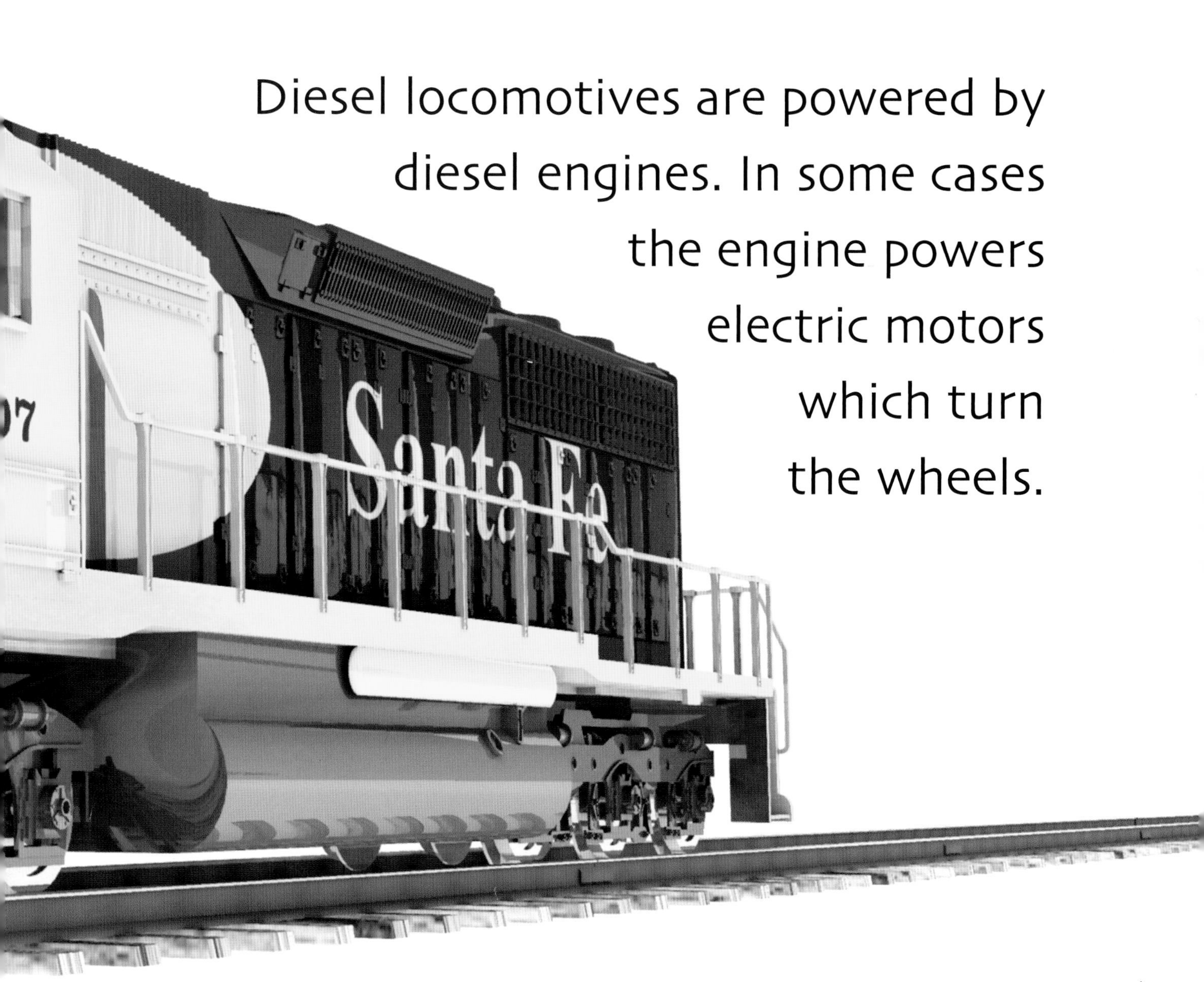

# Electric

Electric locomotives are powered by electricity from a third rail or from overhead lines. Flexible rods called pantographs on the roof of the locomotive contact the overhead wire to transfer the electricity.

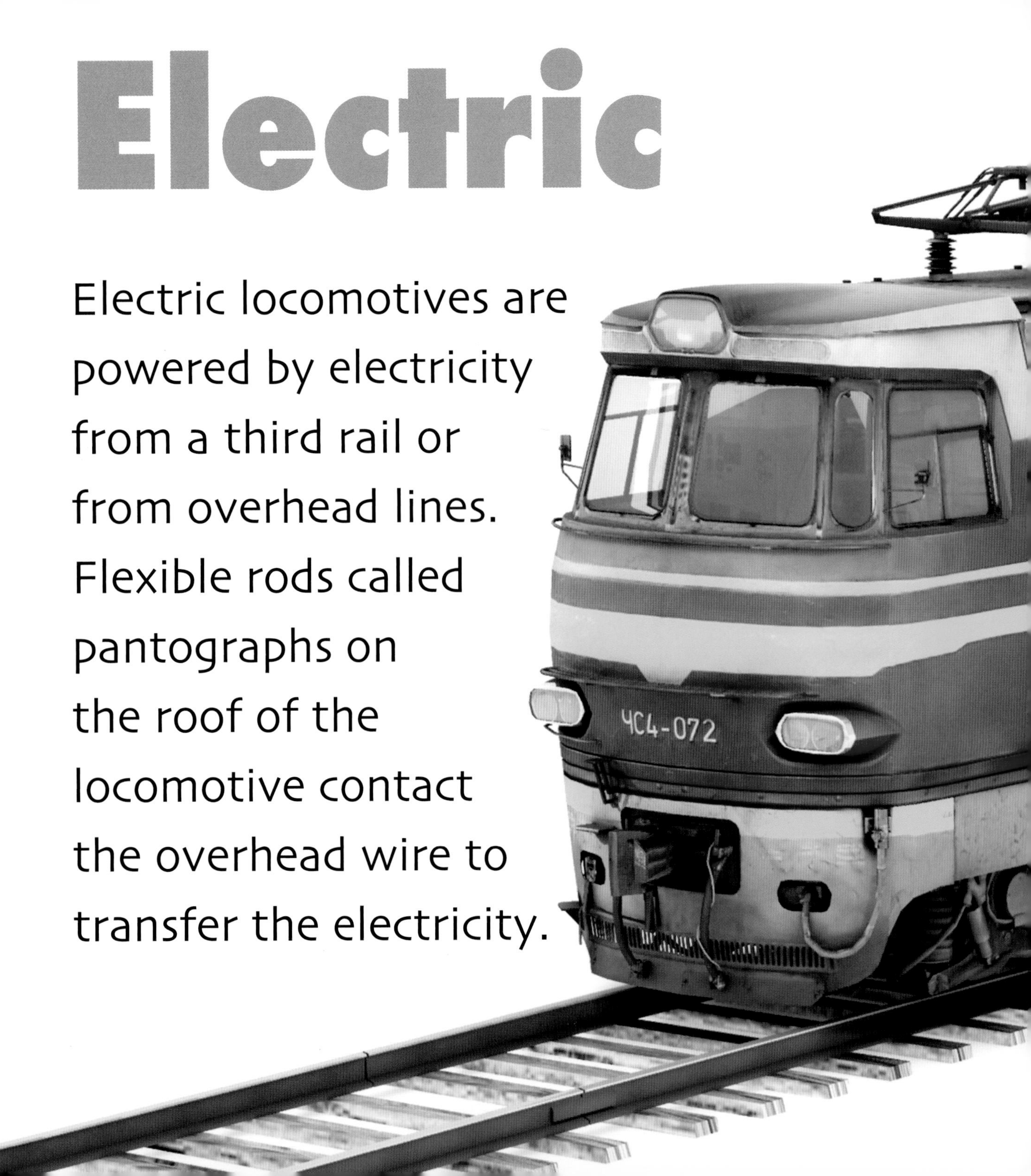

**Mechanic Mike says:**
At the flick of a switch this locomotive can travel forward or backward.

This ChS4-012 pulled passenger carriages between Moscow and Odessa, in Russia.

Although electric locomotives can be fast, this Russian Skoda ChS4's top speed is 99 mph (160 km/h).

The latest version of this locomotive can pull 32 passenger carriages.

Did you know that the ChS4-012 has been retired and can be seen in the Kiev museum of railway transport?

It is powered by electric motors.

# High-Speed

High-speed trains carry passengers on special tracks over long distances at speeds of around 200 mph (320 km/h).

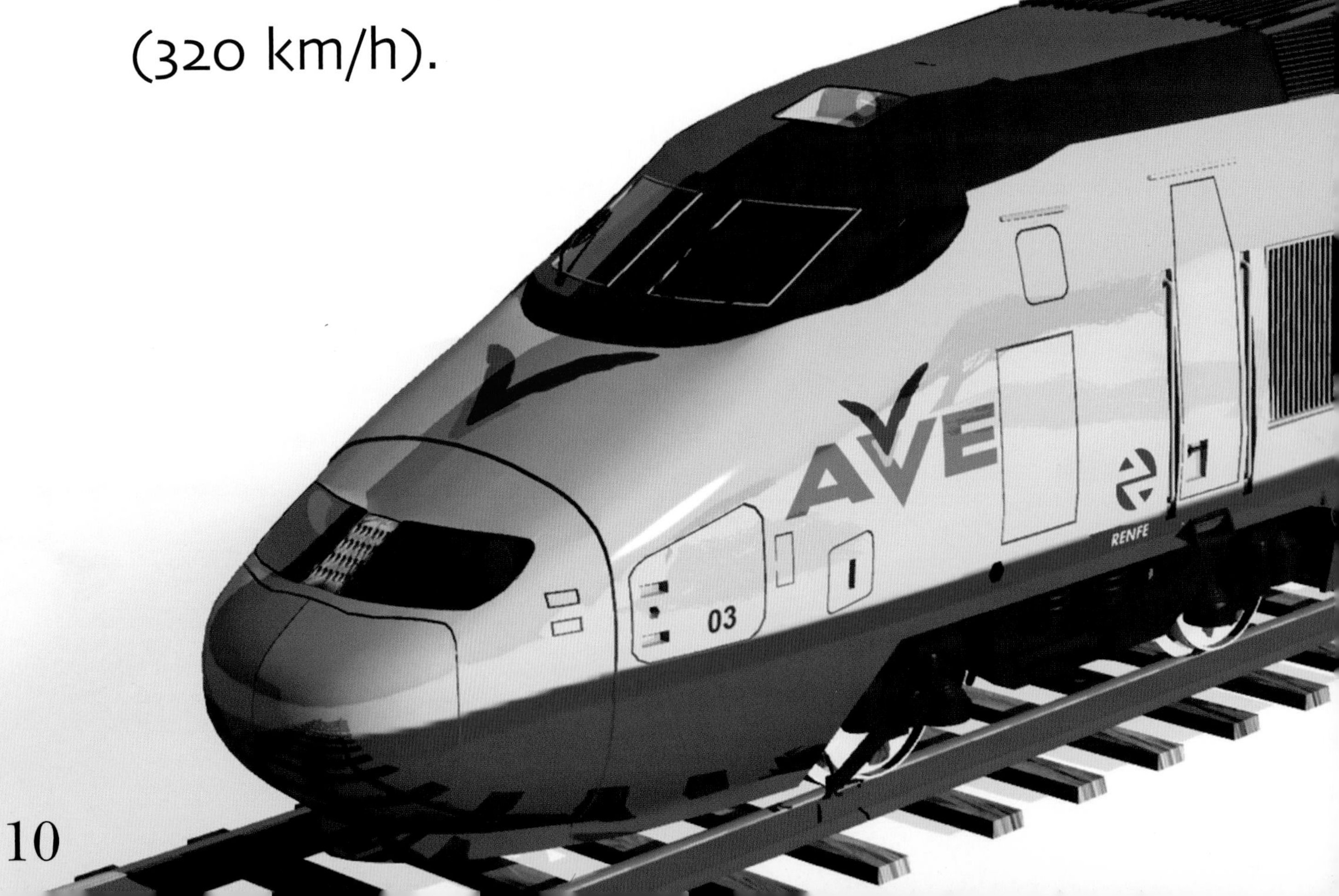

The Chinese CRH380A is the fastest high-speed train in operation. It is designed to travel at 217 mph (350 km/h).

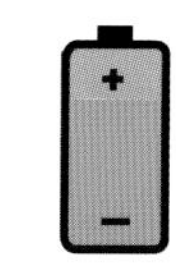

It is powered by electric motors.

High-speed trains require only one driver.

Did you know that the French TGV set a speed record of 357.2 mph (574.8 km/h)?

This Spanish high-speed train can achieve speeds of up to 193 mph (310 km/h) on runs between Madrid and Seville.

**Mechanic Mike says:**

China has the world's longest high-speed rail line, which runs 1,372 miles (2,208 km) from Beijing in the north to Shenzhen on the southern coast.

# Freight

Trains don't just carry people. They are an economical way of transporting all sorts of goods from oil, gas, and coal to grain, cattle, and ore.

**Mechanic Mike says:**
In some countries rolling highway trains are used. These freight trains have special wagons that allow trucks to drive straight onto the train and drive off again when the destination is reached.

Did you know the Daqin Railway in China transports more than 1.1 million tons (1 million tonnes) of coal to the east seashore every day?

Freight trains are usually limited to around 75 mph (120.7 km/h).

Loads can be 145 tons (130 tonnes) per wagon and tens of thousands of tons per train.

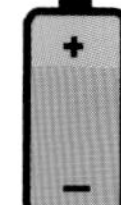

Freight trains are normally pulled by diesel locomotives.

Some freight trains can be over 4.3 miles (7 km) long.

The first **rapid transit system** was the London Underground, which opened in 1863.

Did you know that the busiest rapid transport systems in the world are the Tokyo subway, the Seoul Metropolitan Subway, and the Moscow Metro? The New York City Subway has the record for the most stations.

The train's top speed is 50 mph (80 km/h), while the average speed is 25 mph (40 km/h).

The Danish Copenhagen metro system carries over 137,000 people per day. Each three-carriage train holds up to 96 seated and 204 standing passengers.

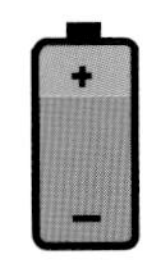

The train is powered by electric motors. The electricity is picked up from a third electrified rail.

# Rapid Transit

Rapid transit trains carry passengers around **urban** areas. The trains run frequently and are designed to carry lots of people. They often travel on lines above roads and underground.

**Mechanic Mike says:**
Some rapid transit trains, like this Copenhagen Metro in Denmark, are completely automated and have no driver.

The Colorado Rail Bilevel rail cars are 19.75 feet (6 m) tall.

Did you know that some countries, such as the UK, don't have bilevel trains since they won't fit under many of the bridges?

Some high-speed bilevel trains can travel over 100 mph (160 km/h).

A four-car set can carry around 400 people.

This is a multiple unit train consisting of self-propelled carriages, using electricity to drive electric motors in many of the carriages.

# Bilevel

**Mechanic Mike says:**
Bilevel trains can be seen in many countries around the world. This one runs in the Netherlands.

These tall trains, also known as double-deckers, have two levels for passengers. This allows them to carry more people in a shorter train. A longer train would need longer platforms to be built.

# Tram

These rail vehicles run on tracks along city streets, and sometimes on special rail lines as well. Most trams today use electrical power, usually supplied by a pantograph.

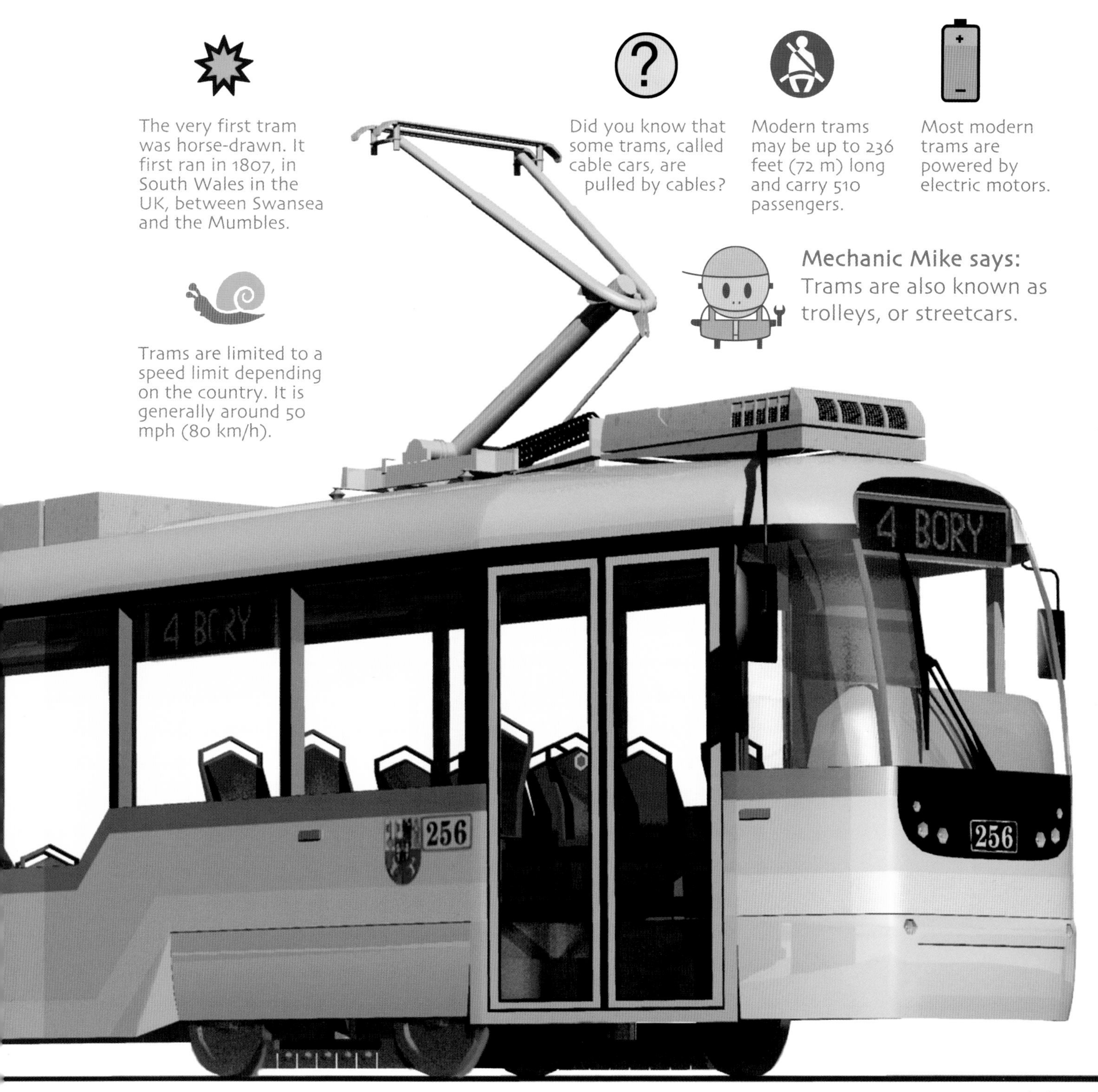


The very first tram was horse-drawn. It first ran in 1807, in South Wales in the UK, between Swansea and the Mumbles.

Did you know that some trams, called cable cars, are pulled by cables?

Modern trams may be up to 236 feet (72 m) long and carry 510 passengers.

Most modern trams are powered by electric motors.

**Mechanic Mike says:**
Trams are also known as trolleys, or streetcars.

Trams are limited to a speed limit depending on the country. It is generally around 50 mph (80 km/h).

Some monorail designs have the trains suspended from a single rail rather than sitting on top of it.

Monorails are generally quite slow. The Tokyo Monorail runs at 50 mph (80 km/h).

The first monorail to carry passengers operated during the 1820s in Hertfordshire, in the UK.

The busiest monorail line is the Tokyo Monorail. It carries more than 300,000 passengers daily.

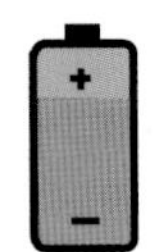

Almost all modern monorails are powered by electric motors.

**Mechanic Mike says:**
Monorails like this one in Las Vegas are popular with tourists, as the height gives them a more **elevated** view of the city.

# Monorail

These trains run along one rail which is usually high above the ground. The train has rubber wheels which can grip the monorail, allowing it to go up and down slopes.

# Maglev

Maglev trains use powerful **electromagnets** to hover above the rail. The magnets are also used to move the train. As there is no contact with the rail, these trains are very smooth, quiet, and fast.

The first commercial maglev train was called "MAGLEV" and opened in 1984 near Birmingham, in the UK.

The highest recorded speed of a maglev train is 361 mph (581 km/h).

The top operational speed of this Shanghai Maglev train is 268 mph (431 km/h), making it the world's fastest train in regular use.

The Shanghai Maglev can carry 244 people.

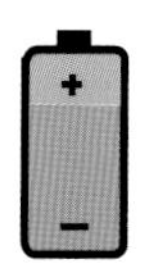

It uses electromagnetic propulsion.

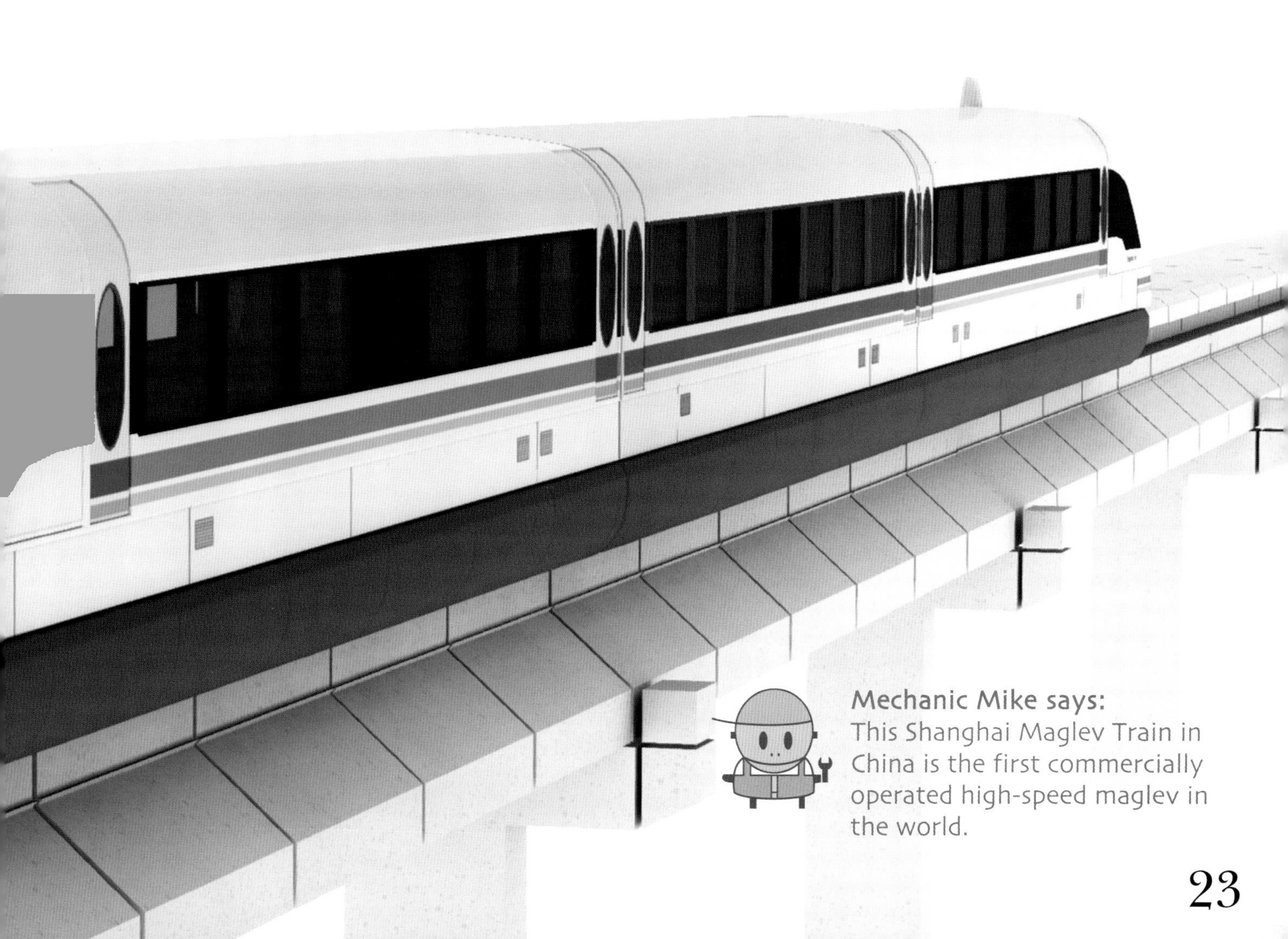

**Mechanic Mike says:**
This Shanghai Maglev Train in China is the first commercially operated high-speed maglev in the world.

# Glossary

**diesel**
A fuel like gasoline used in diesel engines. Diesel engines can be very powerful.

**electromagnet**
A magnet produced using electricity.

**elevated**
Higher than the surrounding area.

**locomotive**
The railway vehicle that pulls a train.

**rapid transit system**
A rail passenger transport system in an urban area.

**urban**
A built-up area such as a city or town.

# Index